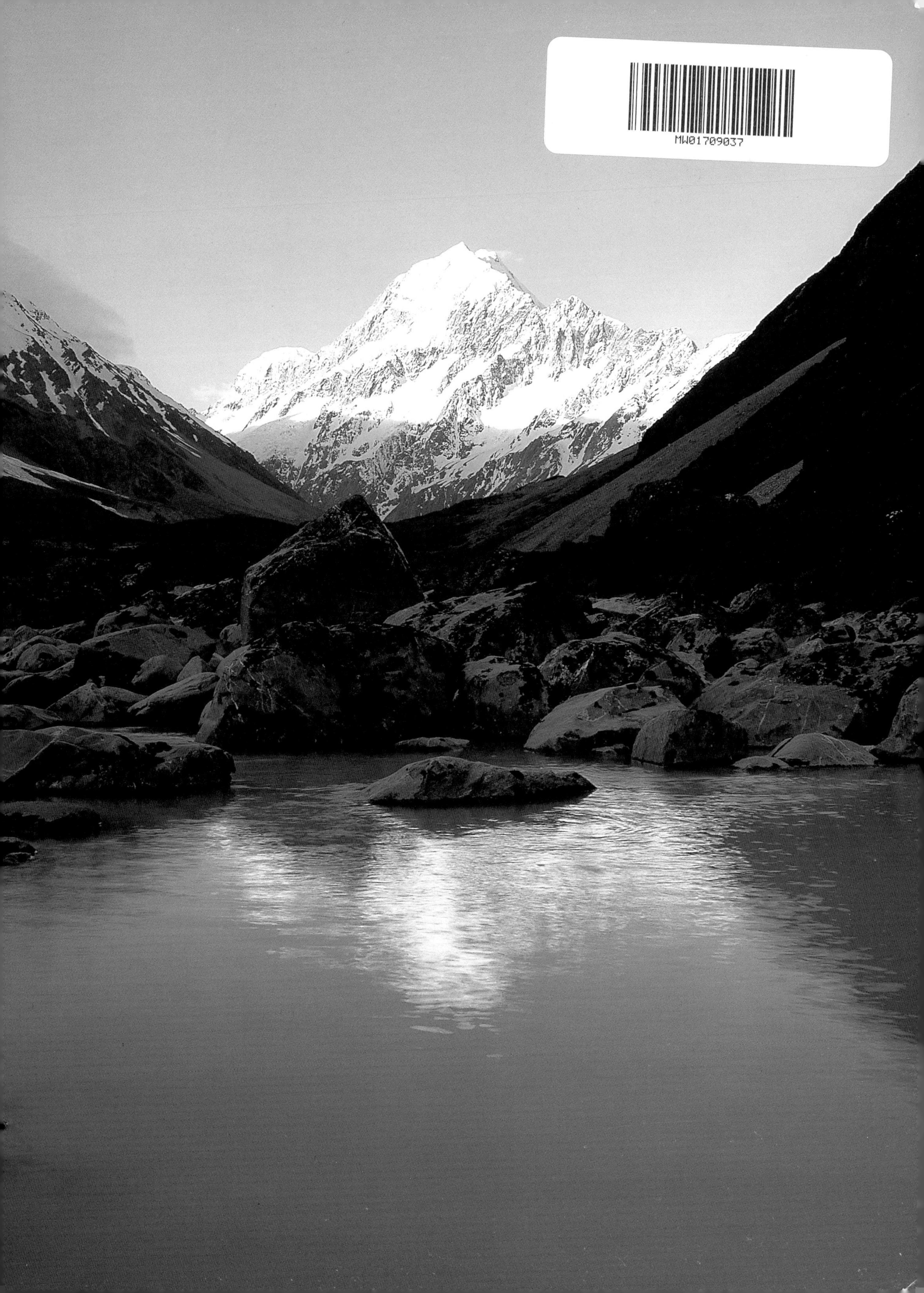

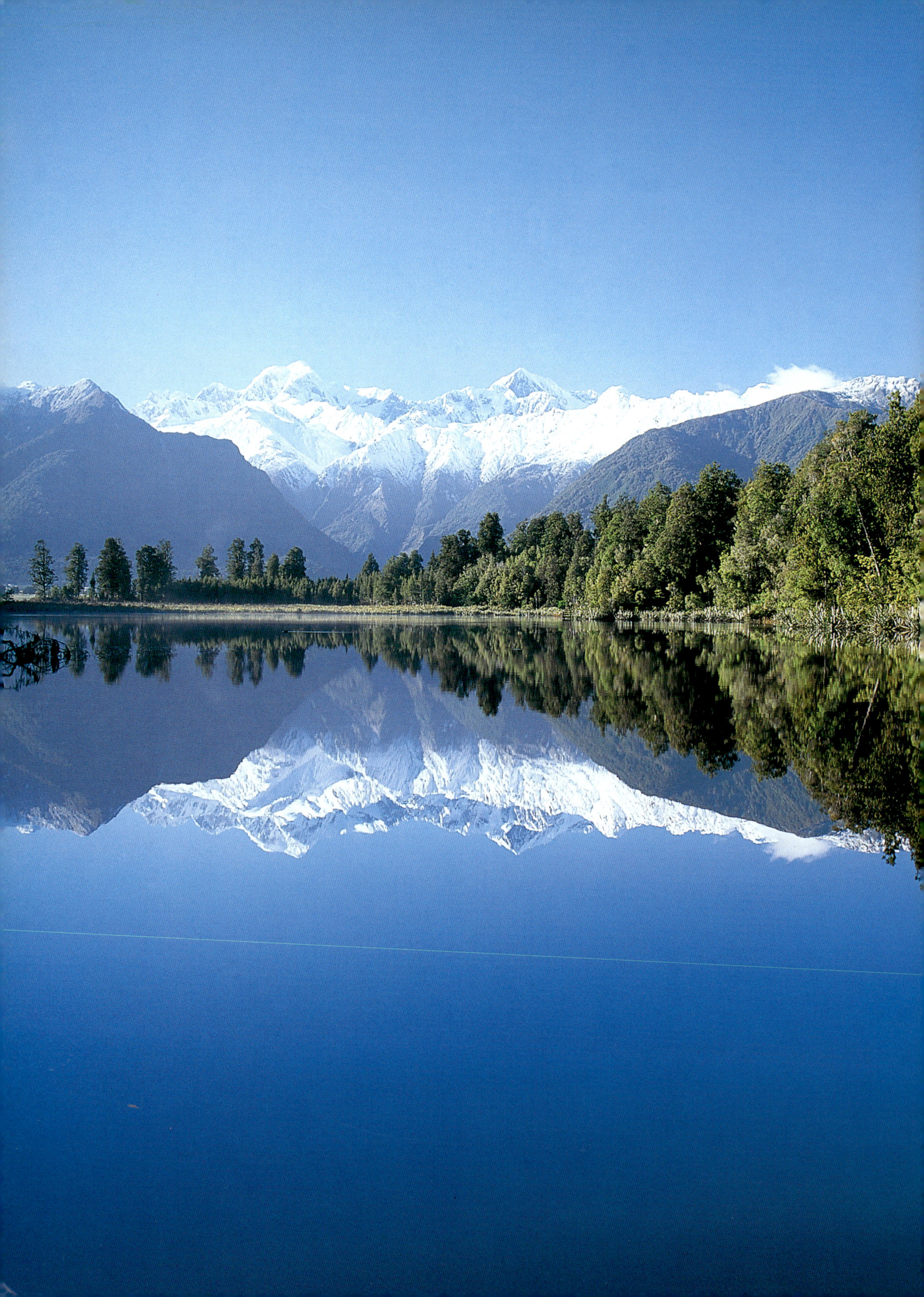

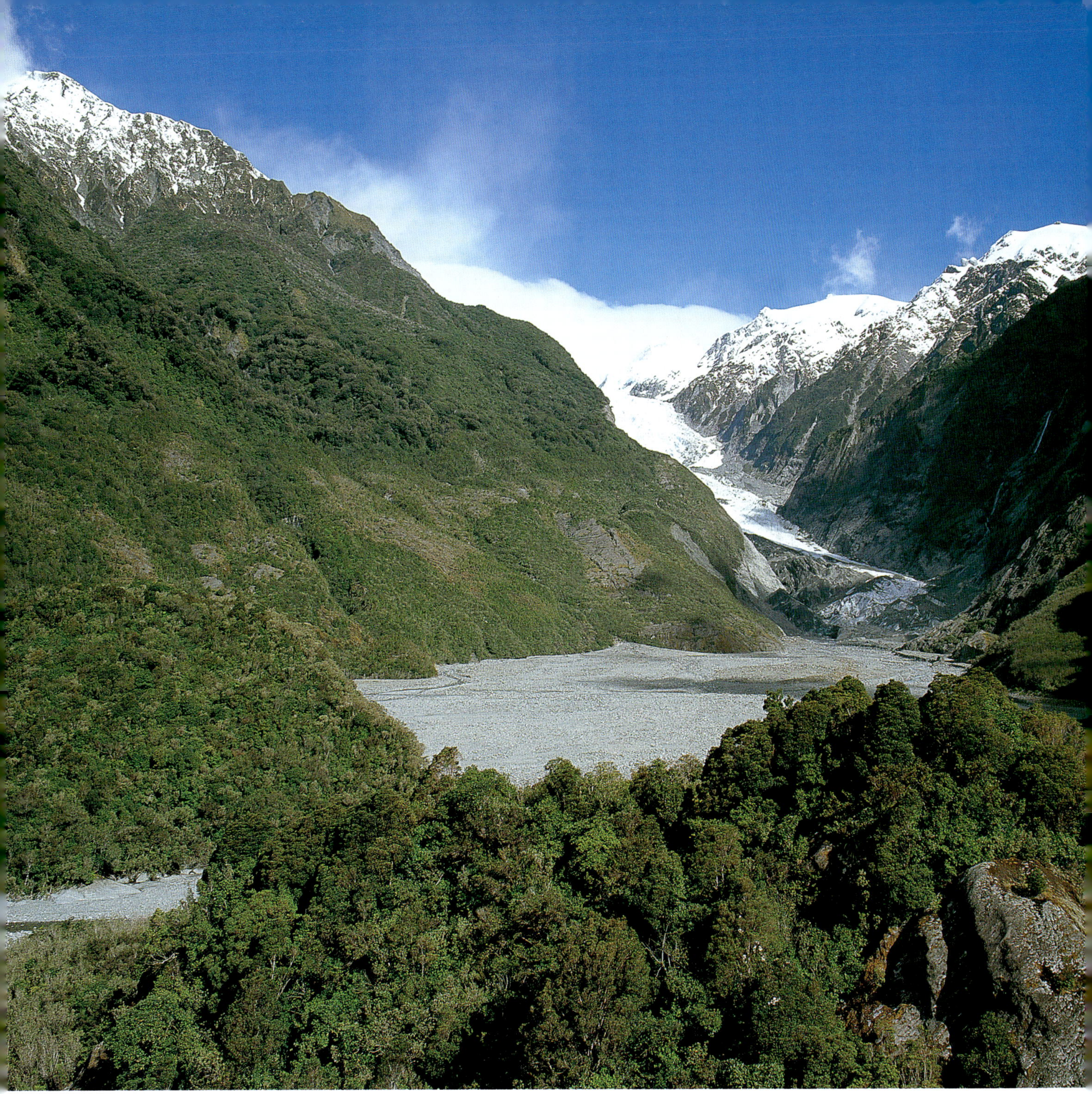

Opposite: Lake Matheson, an ice-melt relic of ancient Fox Glacier, is well known for its calm reflection of Aoraki/Mt Cook, Mt Tasman and the native forests of Westland. In fact, so undisturbed is its surface in this image that it could be upside down!

Above: From Sentinel Rock, in the valley of the Waiho River, it is easy to appreciate the complete course of Franz Josef Glacier from its birth on the slopes of Mts Cook and Tasman to the river that drains its melted ice westward to the Tasman Sea. Throughout the millennia, both Fox and Franz Josef Glaciers have advanced and receded according to climatic conditions. There have been times when their terminal faces have been embraced by temperate rain-forests virtually at sea level.

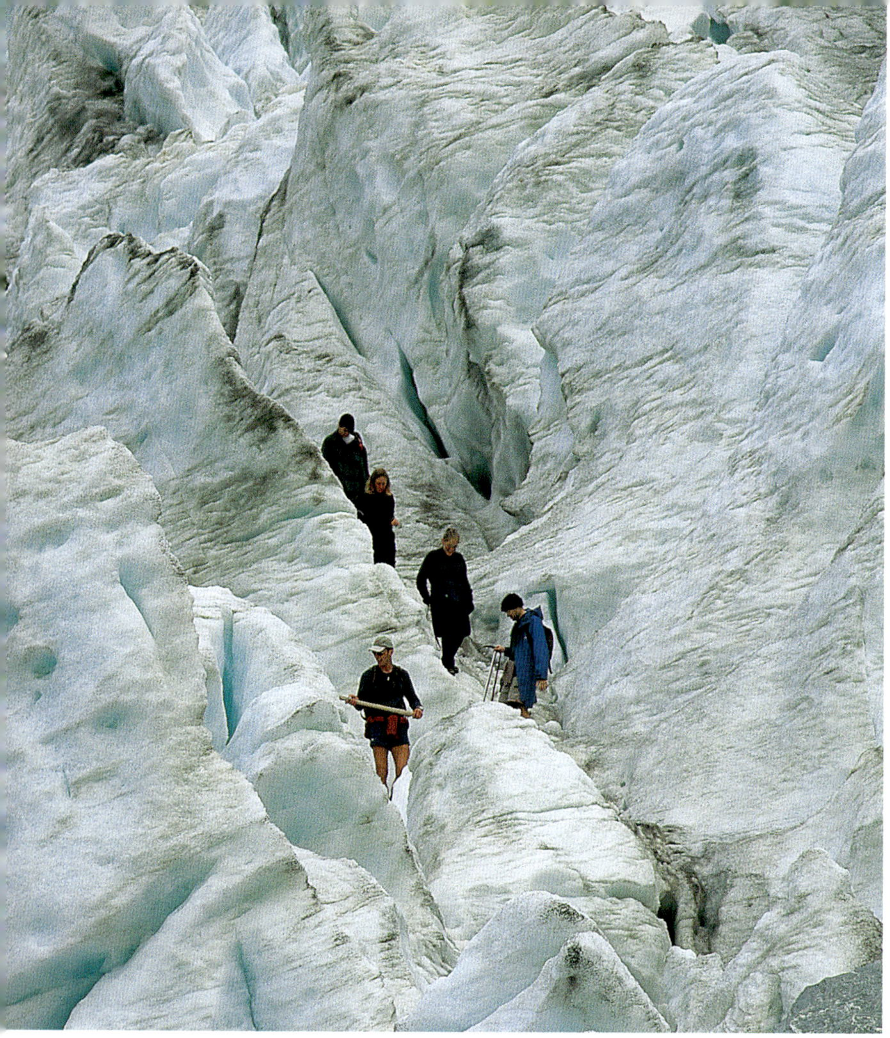

Left: Preferably in the company of an experienced guide it's possible, as this group is doing, to climb the buttresses of Fox Glacier's base and to look into the eerie blue-green depths of its crevasses, formed as the glacier makes a stately progress of about 1.5 metres a day.

Below: The terminal face of Fox Glacier is an untidy clutter of rotting ice and the rock and shingle gouged from the valley of the glacier's own making. Here ice begins to crumble and fall, like calving icebergs, to be carried away to the sea by way of the Fox River.

Opposite: Well bred Hereford cattle graze the lush green pastures of Westland oblivious to the majesty of Fox Glacier and the heights of Douglas Peak (3081 m) and Mt. Haidinger (3066 m) beyond.

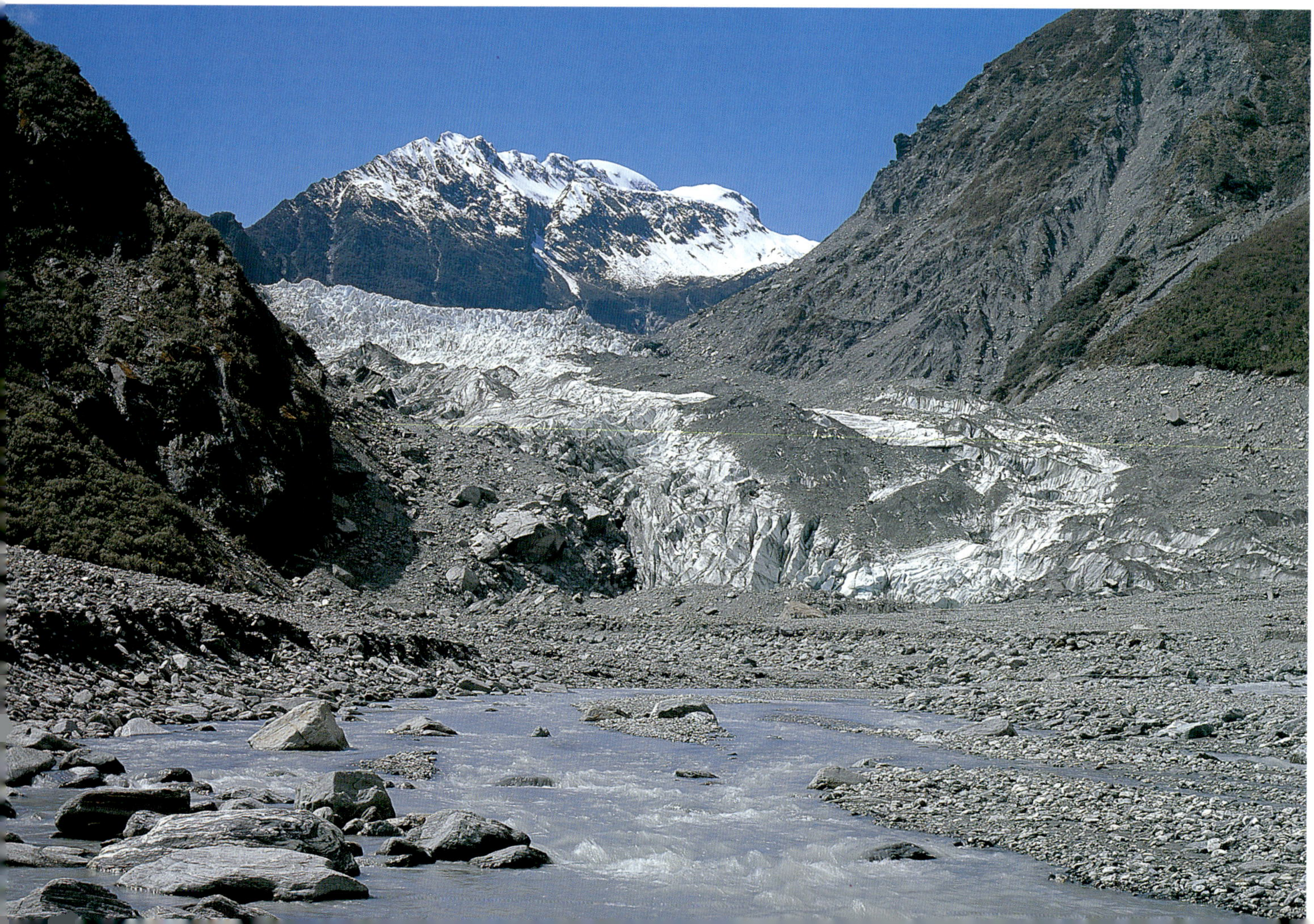

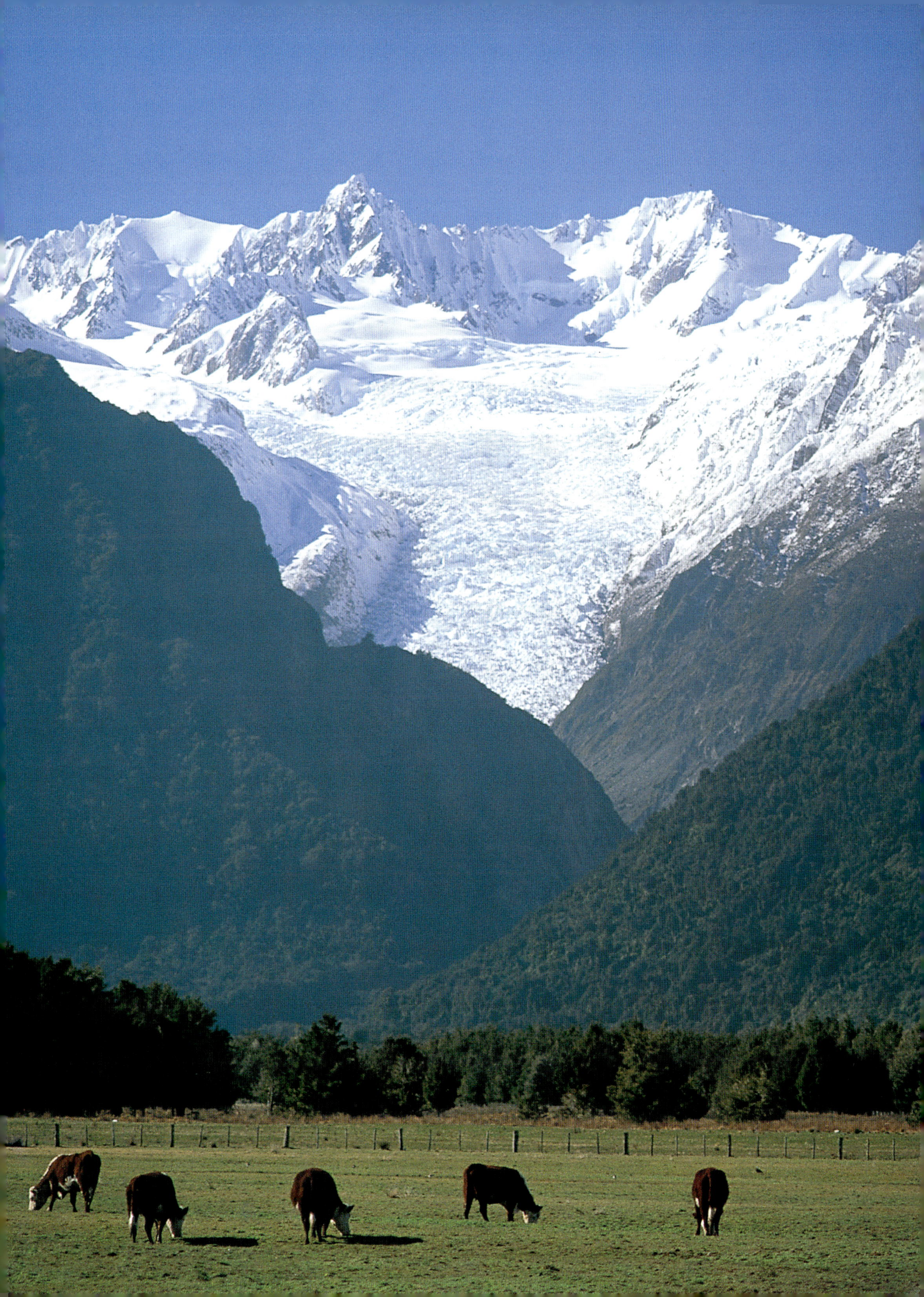

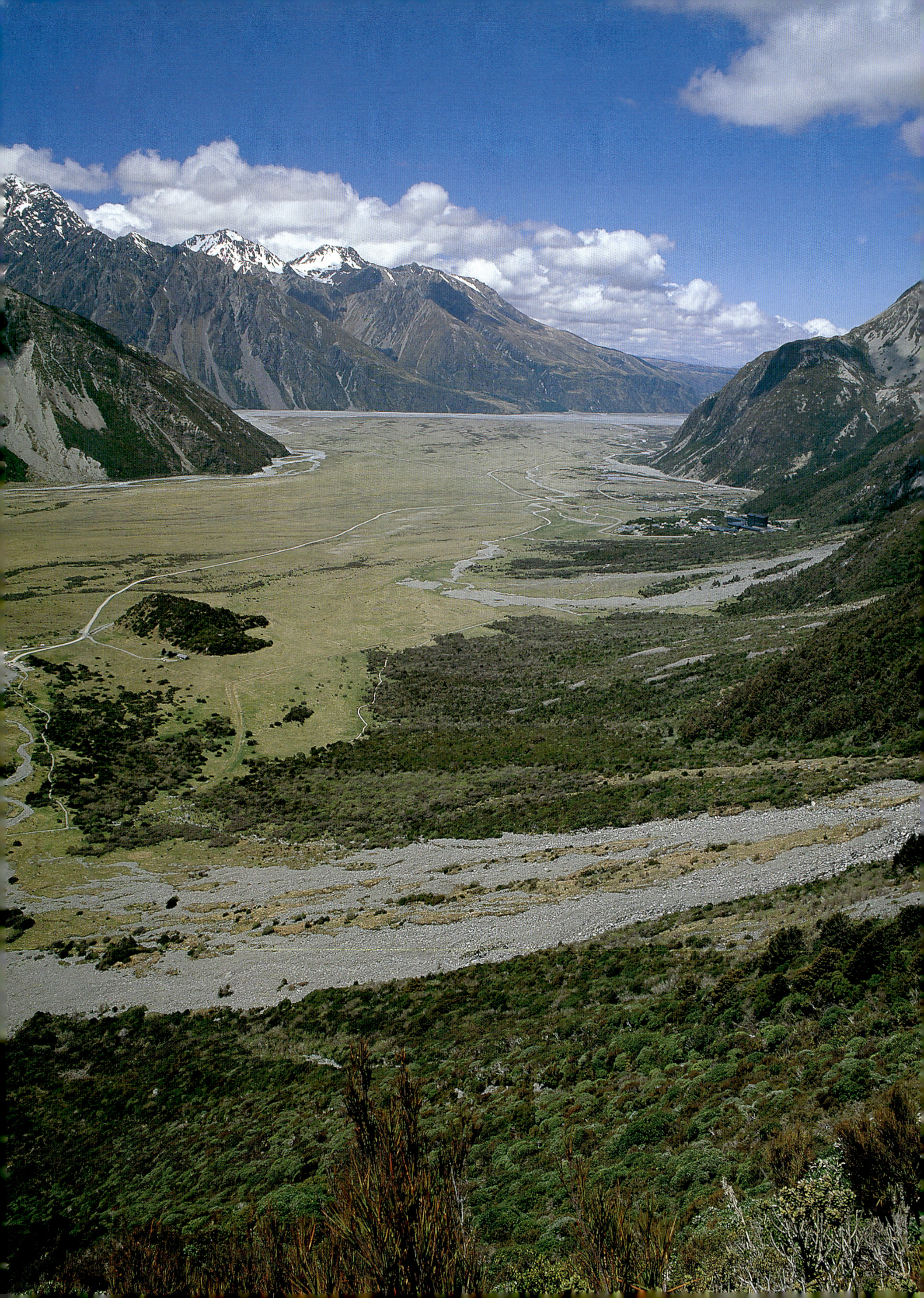

Opposite: Half way up the slopes of the Sealy Range, near the Mueller Glacier, are the small mountain lakes known as Sealy Tarns. From here Aoraki/Mt Cook village and its Hermitage Hotel in the middle distance seem insignificant in the broad, shingle valleys of the Hooker and Tasman Rivers.

Right: From a lookout a short walk north of the village visitors view the Mueller glacial valley and the lake formed at its lower end.

Below: The Mueller and Hooker Glaciers join to form Mueller Lake whose outlet is the Hooker River. To the left is Mt Sefton (3157 m).

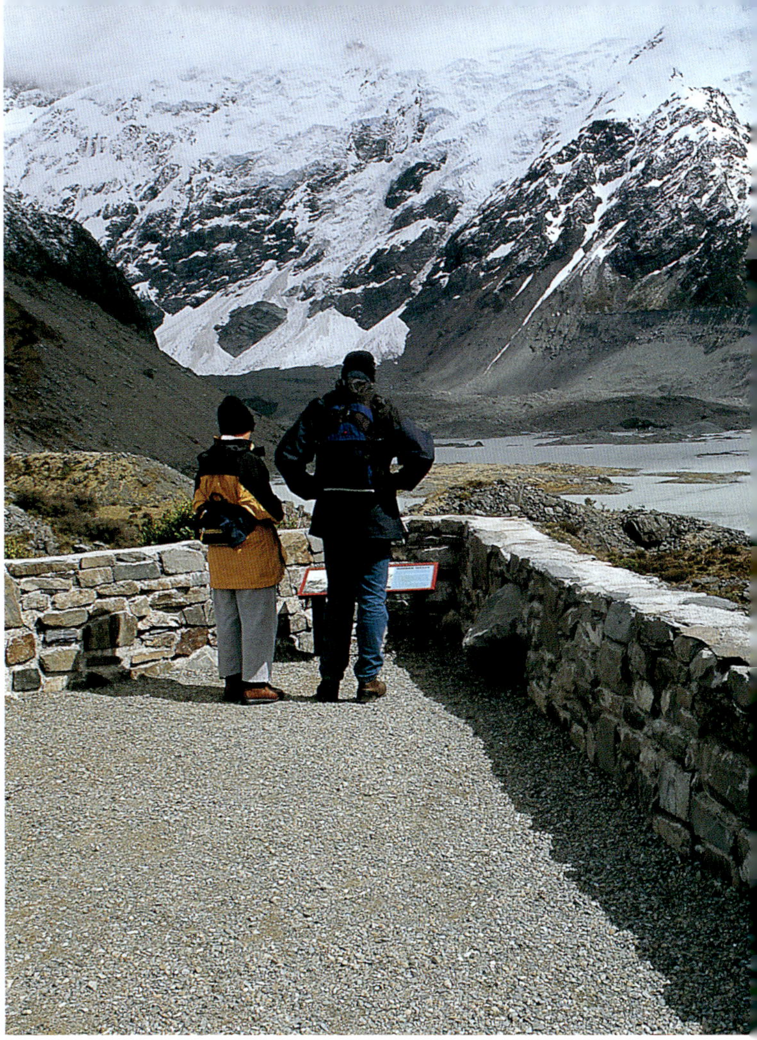

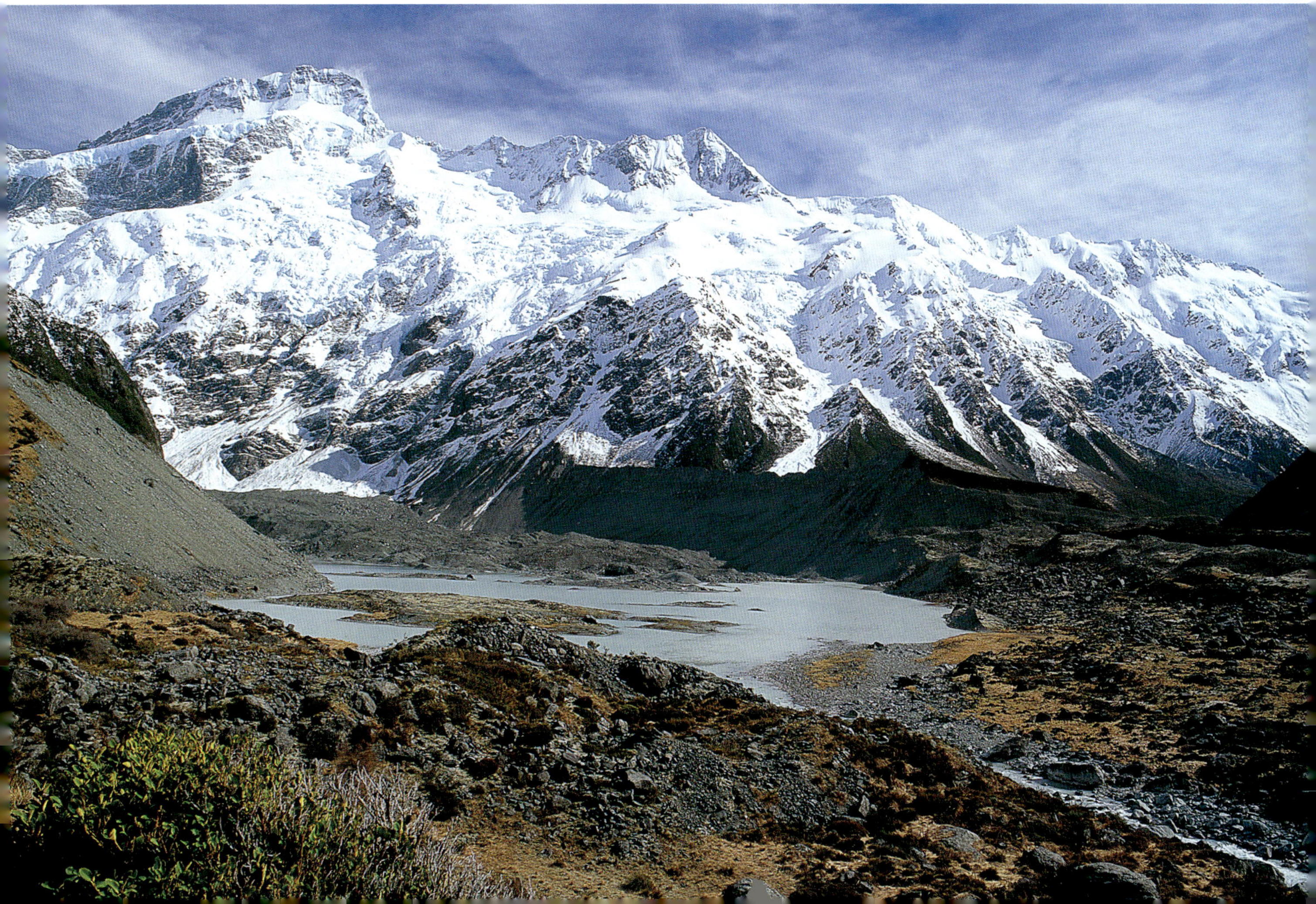

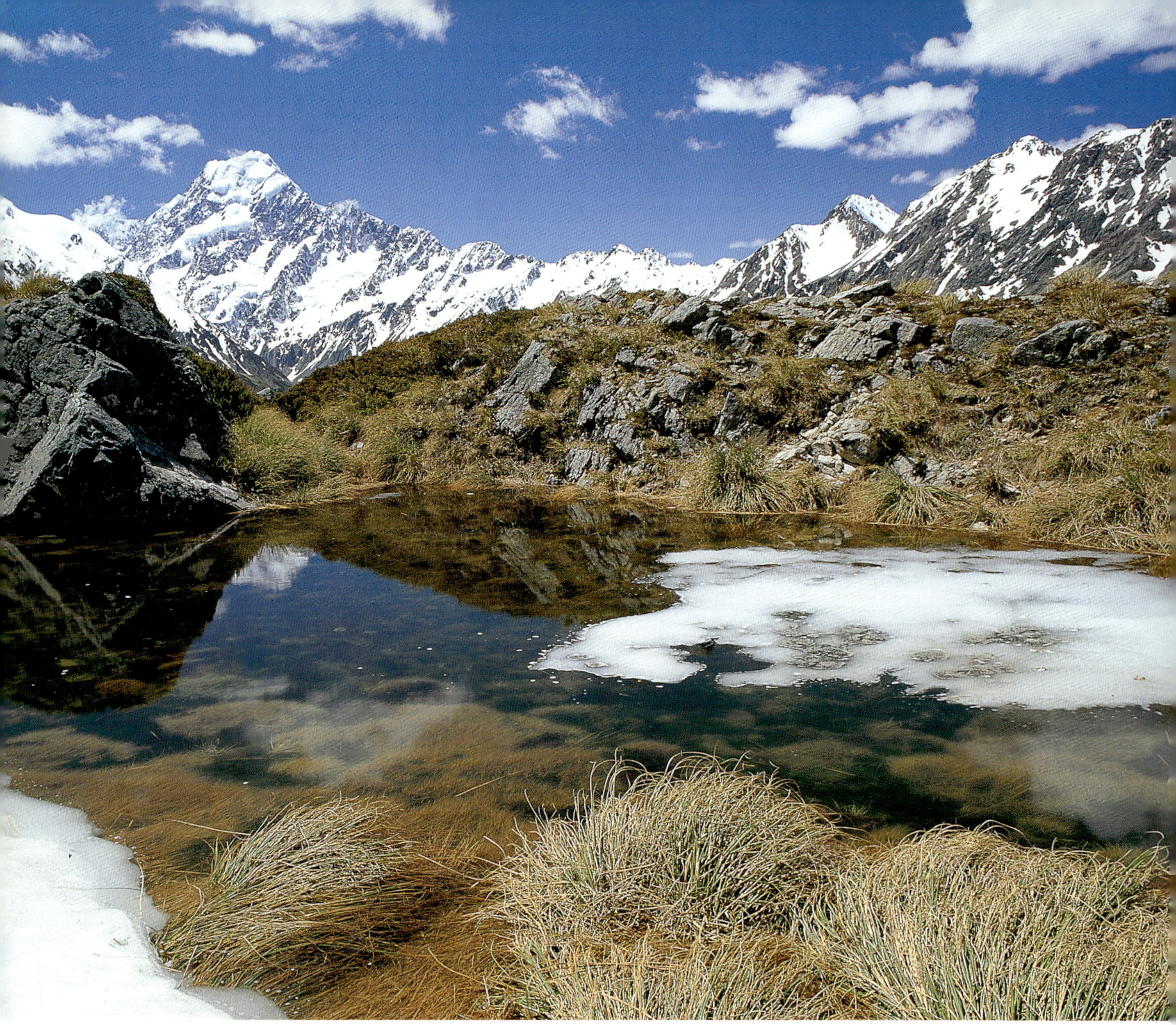

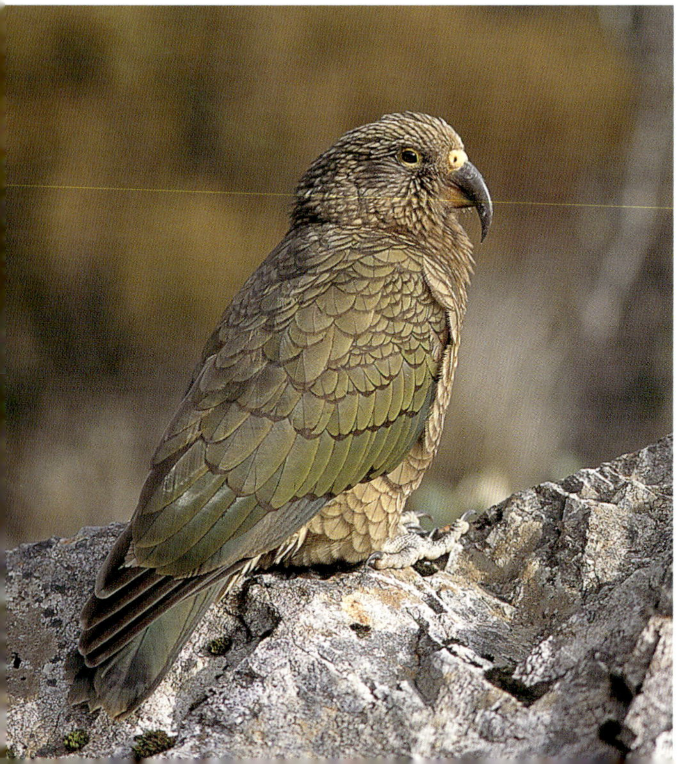

Above: The Sealy Tarns, a 2.5 hr walk from the Aoraki/Mt Cook visitors' centre, are long, narrow ponds formed by melting glacial ice. In summer the water is warm enough to bathe in.

Left: Kea, the fearless native parrot, is a mixture of clown and demon. It's dressed in dull, olive green plumage but the undersides of its wings are an astonishing scarlet. Named by Maori for its raucous cry it struts like an operatic pirate and will steal your lunch, sunglasses, gold watch or passport if given the chance. Beware!

Opposite above: The Sealy Tarns track diverts up a stair climb from the path between Mt Cook village and Kea Point, above the Mueller Glacier.

Opposite below: The forbidding southern faces of Aoraki/Mt Cook seen from the Sealy Range above Mueller Glacier.

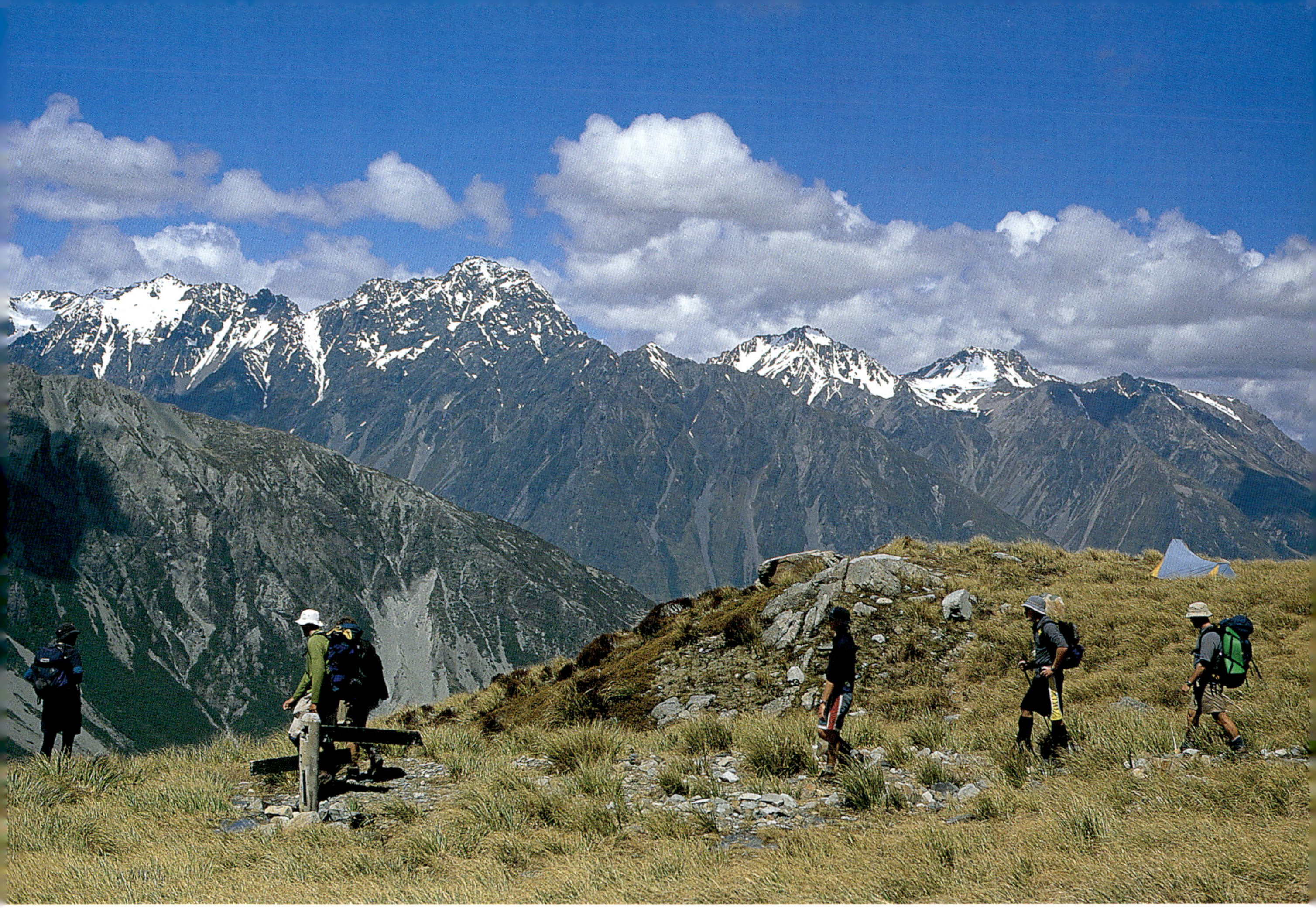

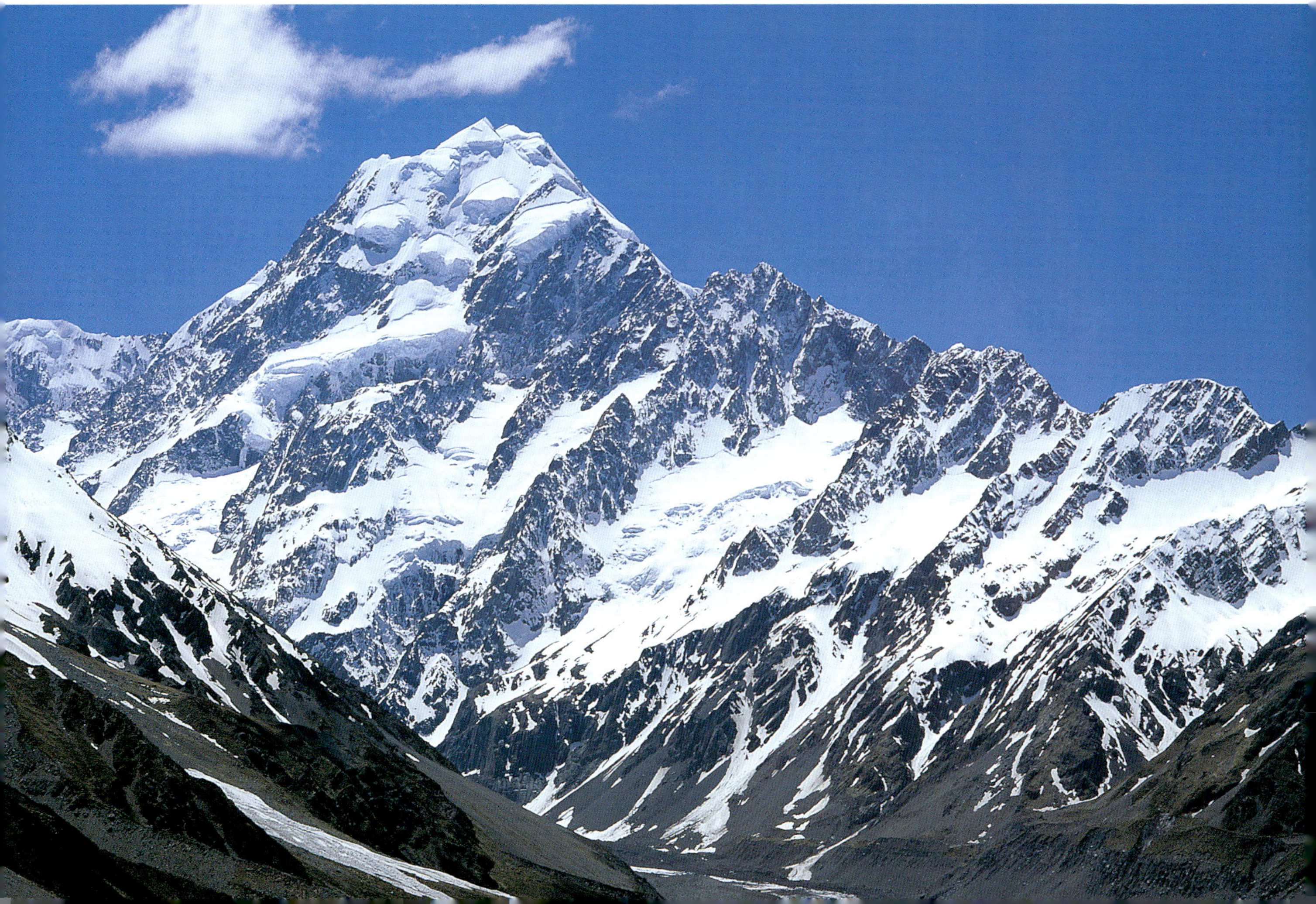

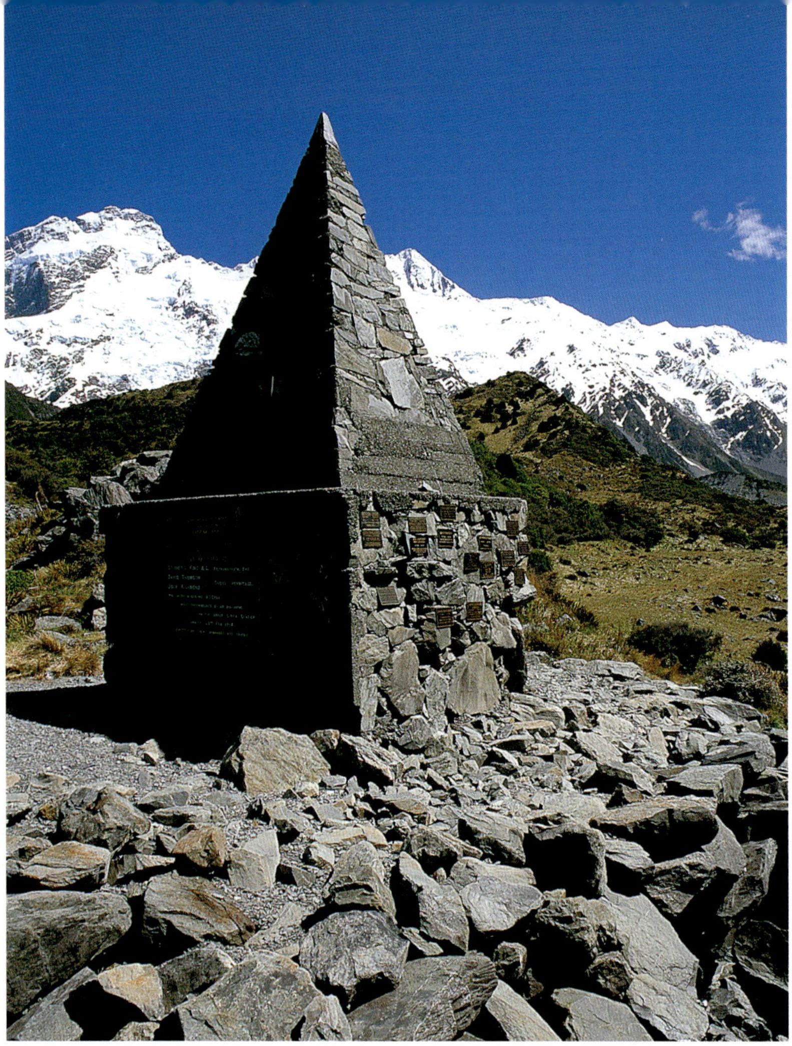

Left: This alpine memorial near the Hermitage Hotel was raised in memory of two guides and their client who were killed in an avalanche on the Linda Glacier in 1914. When the centenary of alpine climbing was celebrated in 1994 it was re-dedicated to all mountaineers and guides who have perished in these mountains.

Below: When the lower ice and snows of Tasman Glacier are melted, glacial scouring is revealed as untidy piles of moraine débris.

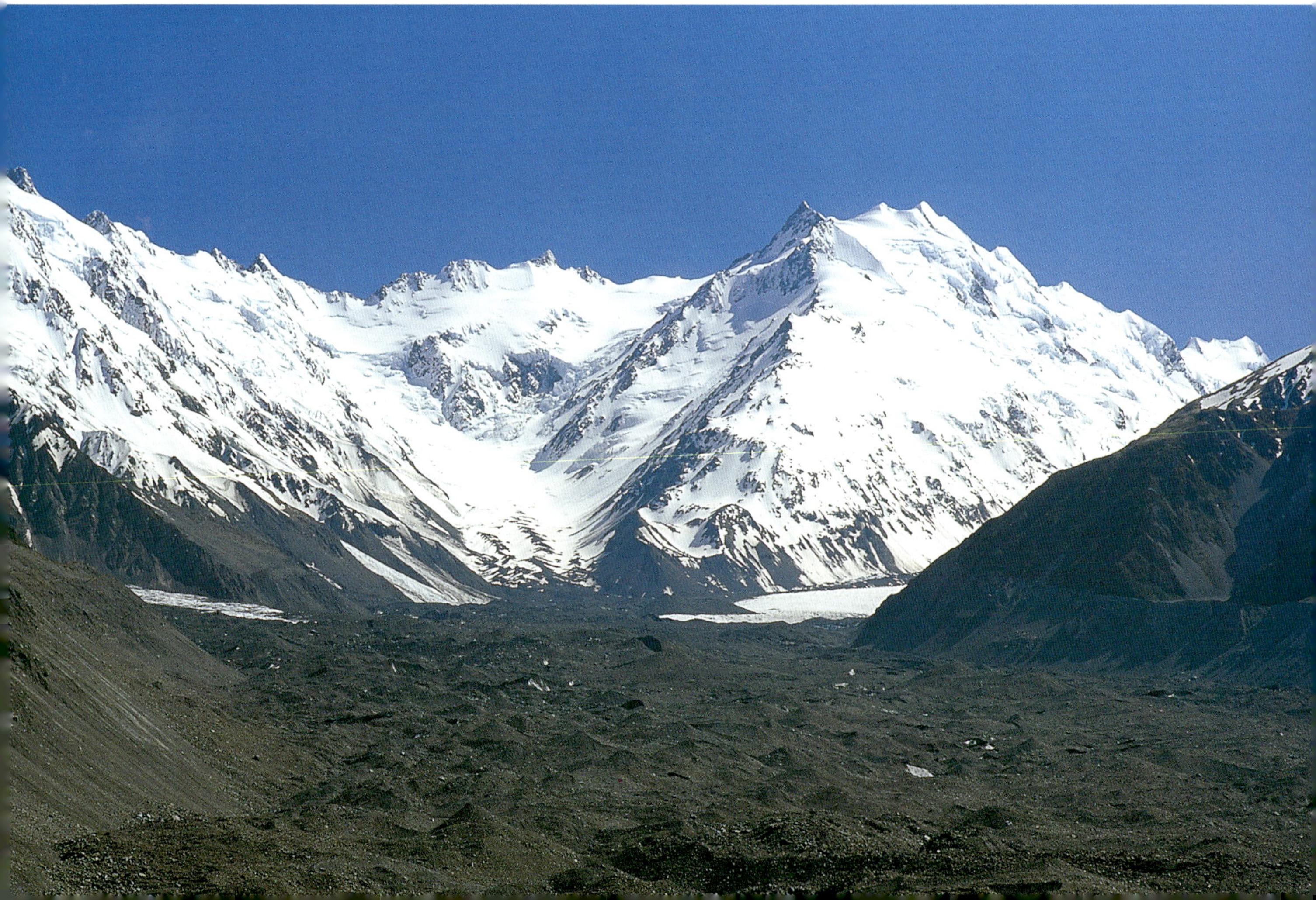

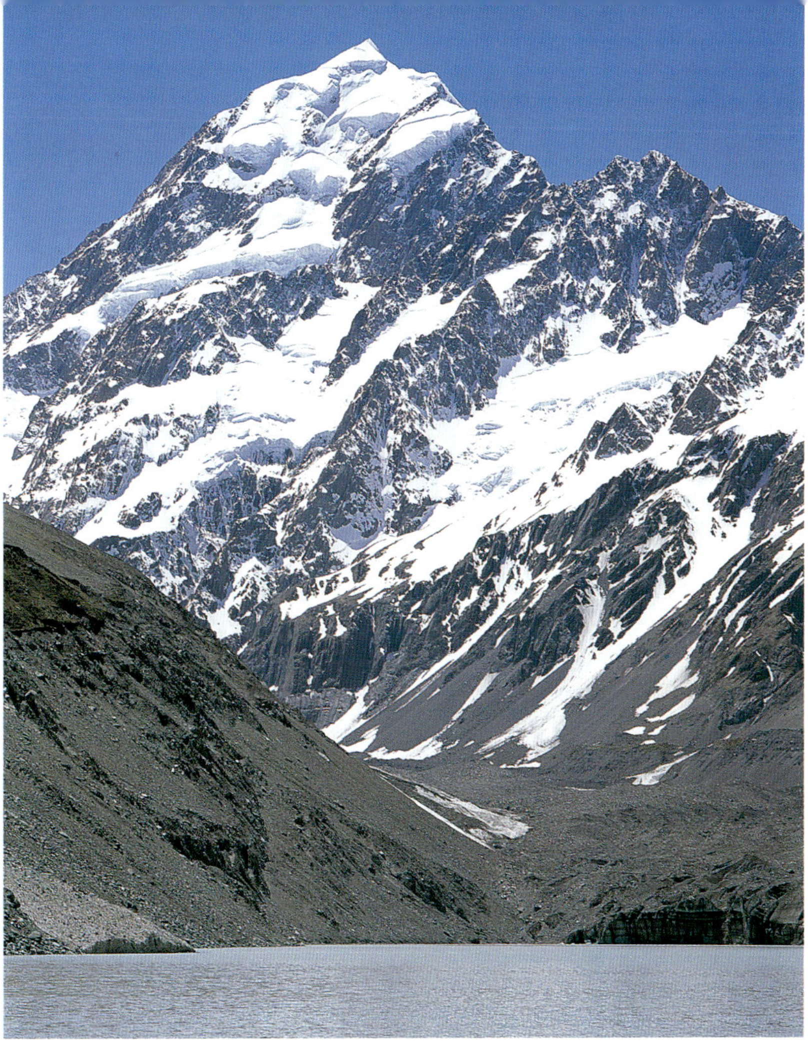

Right: The icy waters of Hooker Lake glitter below the south face of Aoraki/Mt Cook from whose upper-pitch snow fields seem poised to slump and avalanche at any moment.

Below: Who knows how ancient the crystals of this miniature iceberg might be as it floats, tired and gravel-stained in Hooker Lake, below its glacier? It will crumble and melt to join the waters of the Hooker River as they tumble south towards Lake Pukaki.

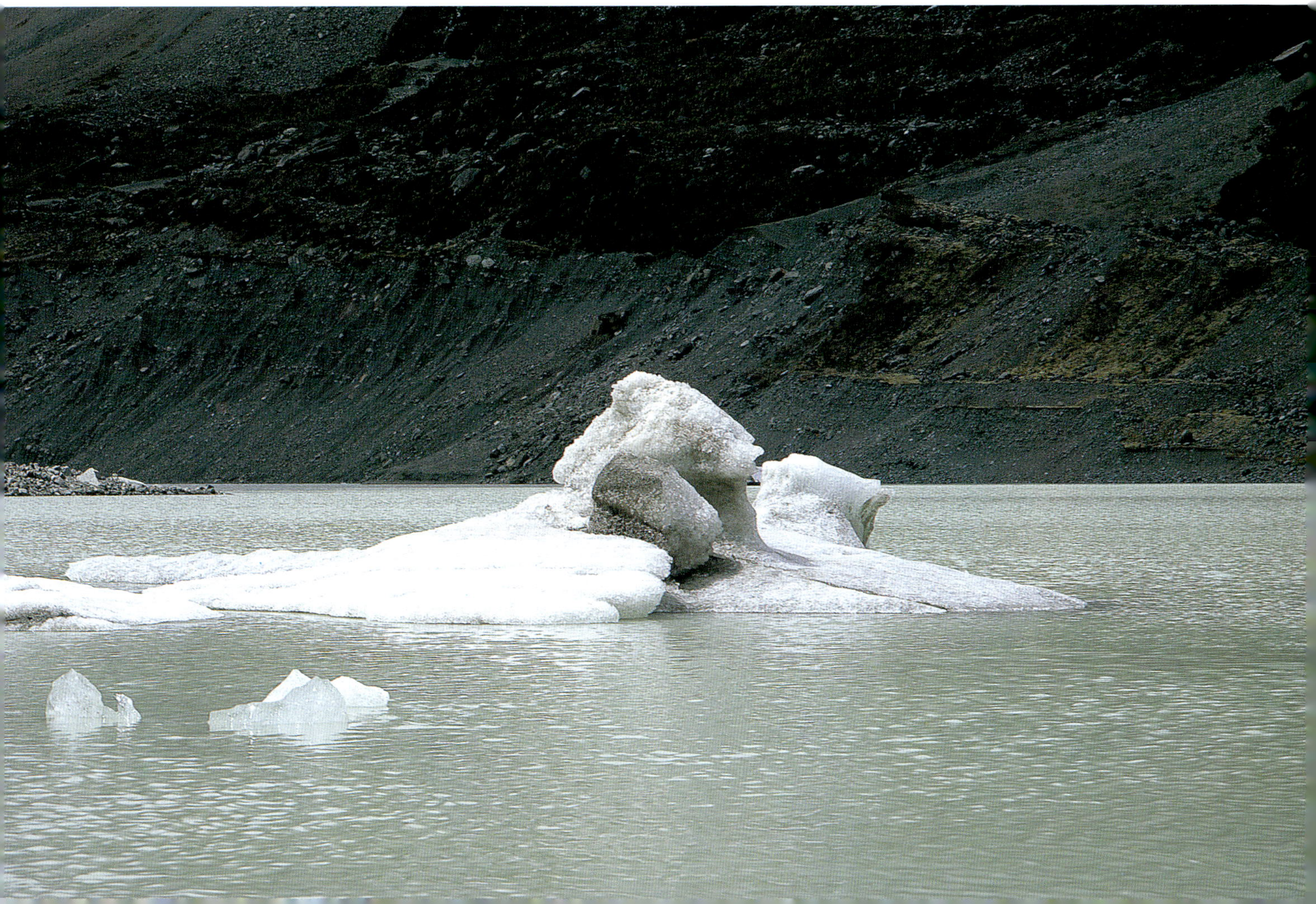

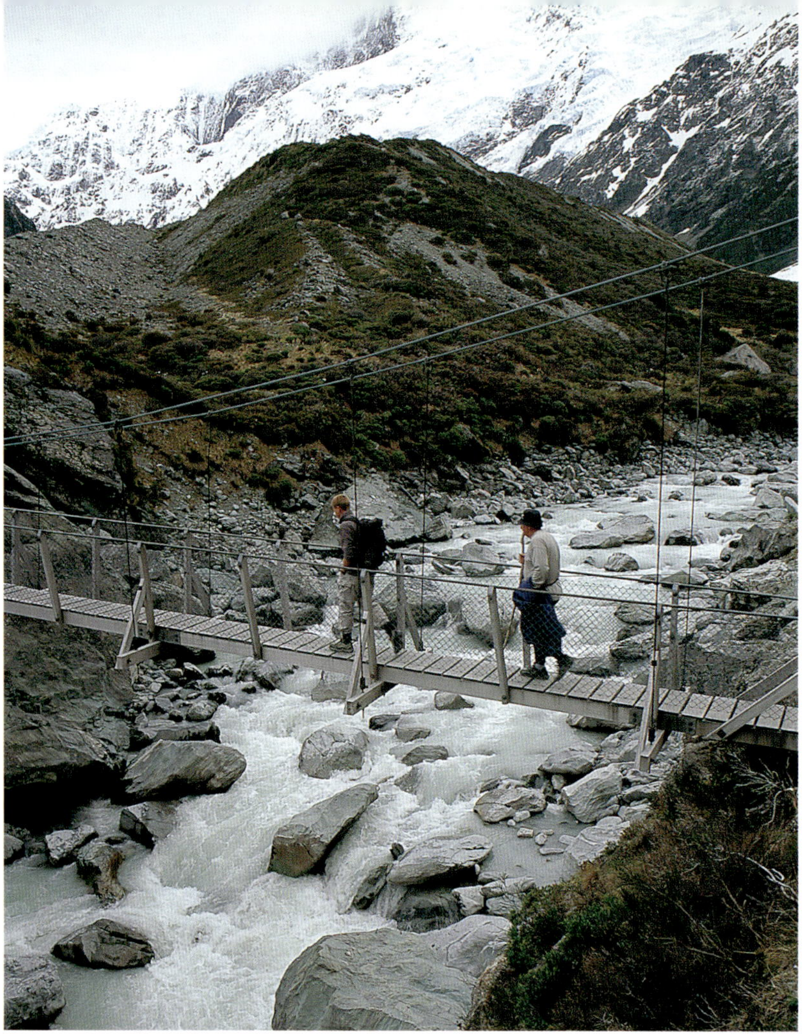

Left: As the morning sun breaks through cloud on distant slopes trampers, still in early shadow, cross the swing bridge over the Hooker River on their way to Hooker Lake. Soon they too will be in sunshine, warming from their exertions.

Below: Stocking Stream Shelter, below Mt Wakefield (2050 m), on the way to Hooker Glacier, is one of 18 huts strategically dotted around the national park. This picture shows not only an impressive variety of alpine plants – a summer garden – but also the shingle 'fans' which are formed by débris deposited by mountain streams.

Opposite: Of all the hundreds of native flowering plants that grow in the valleys the Mount Cook Lily is the most outstanding. Its scientific name *Ranunculus lyallii* informs us that it is not a lily at all; it is, in fact, the world's largest buttercup.

Overleaf: While originating from the western side of the American Rockies, the Douglas fir is in no way out of place among the tussocks of Canterbury below Aoraki/Mt Cook.

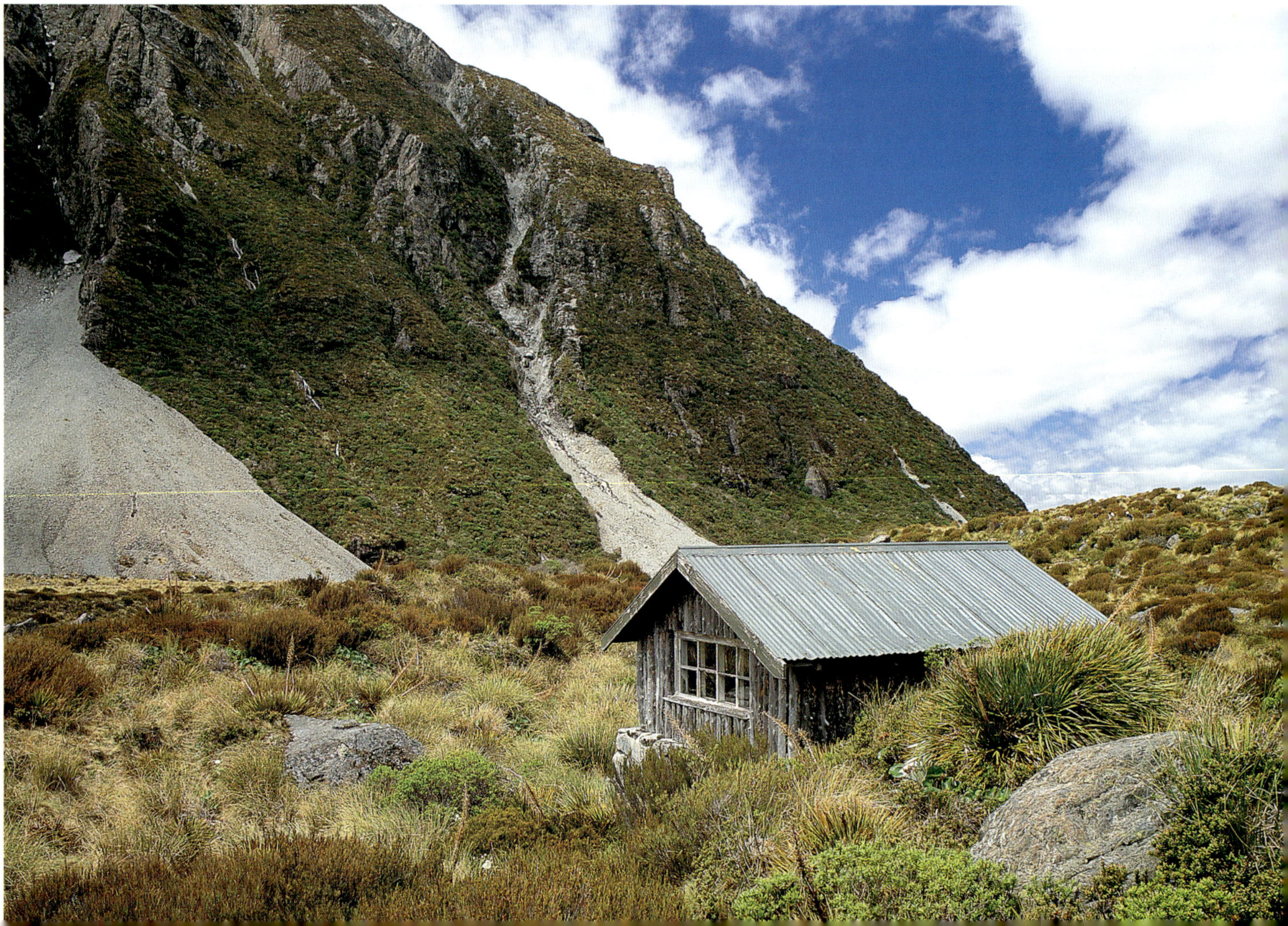

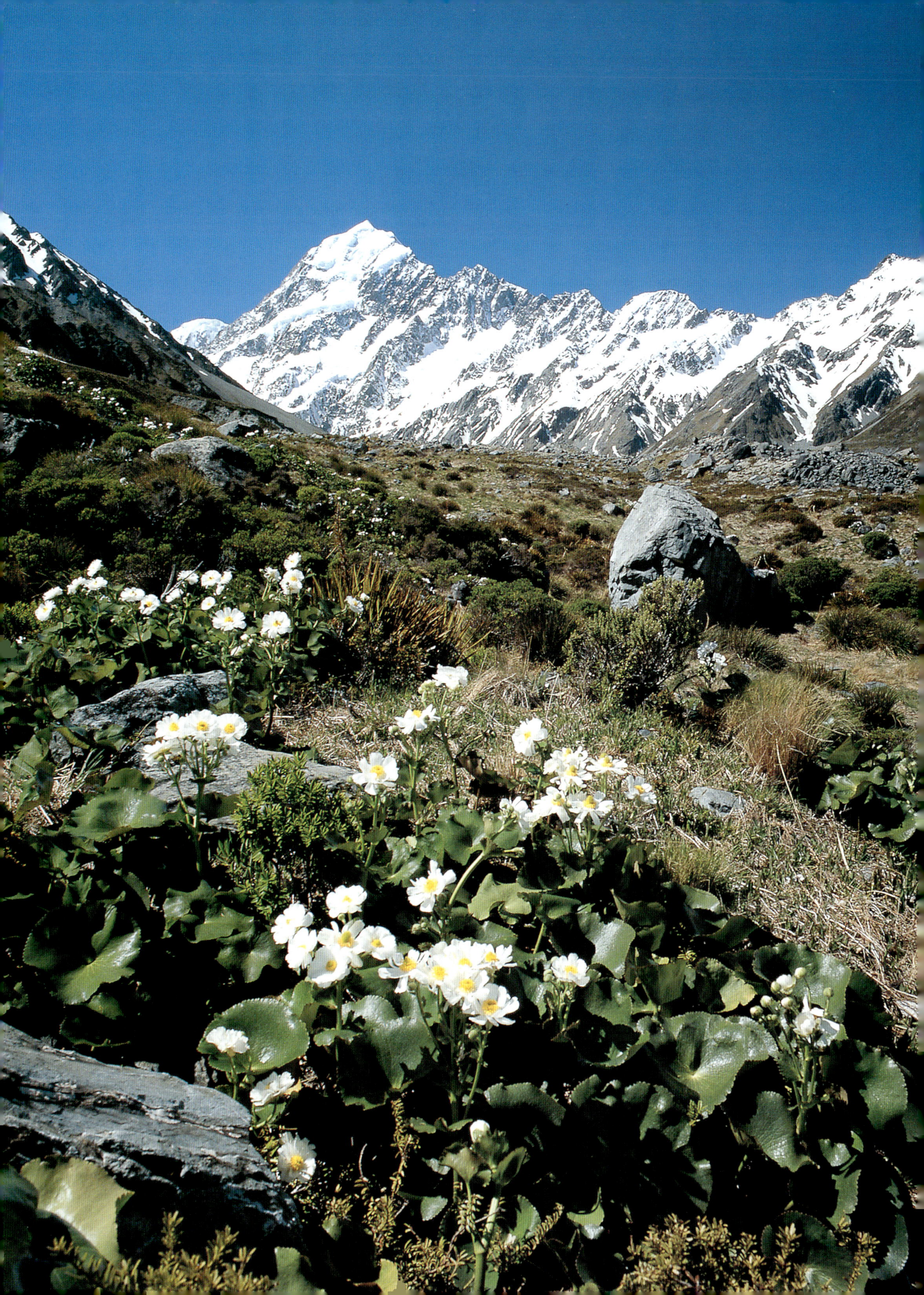

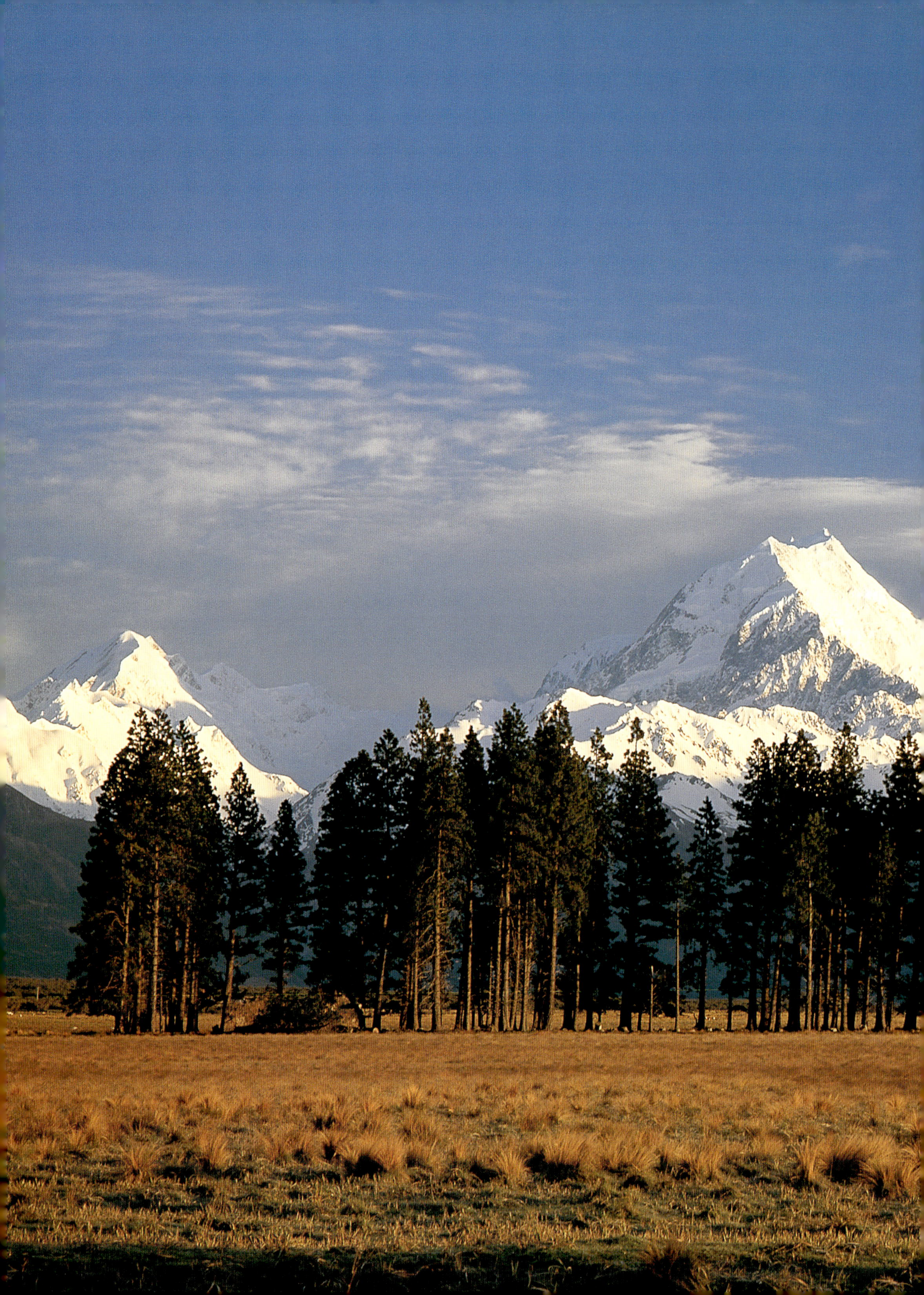

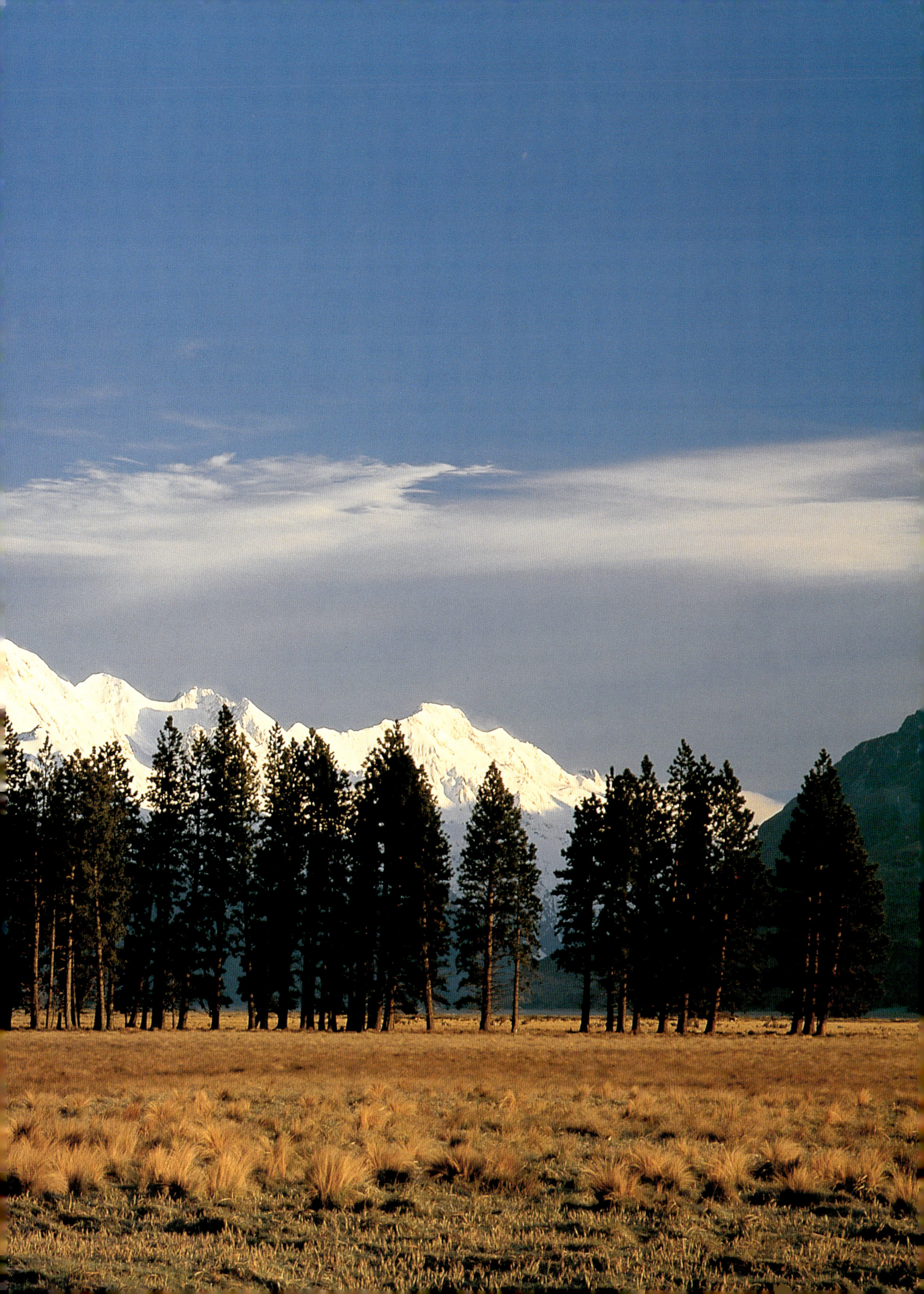

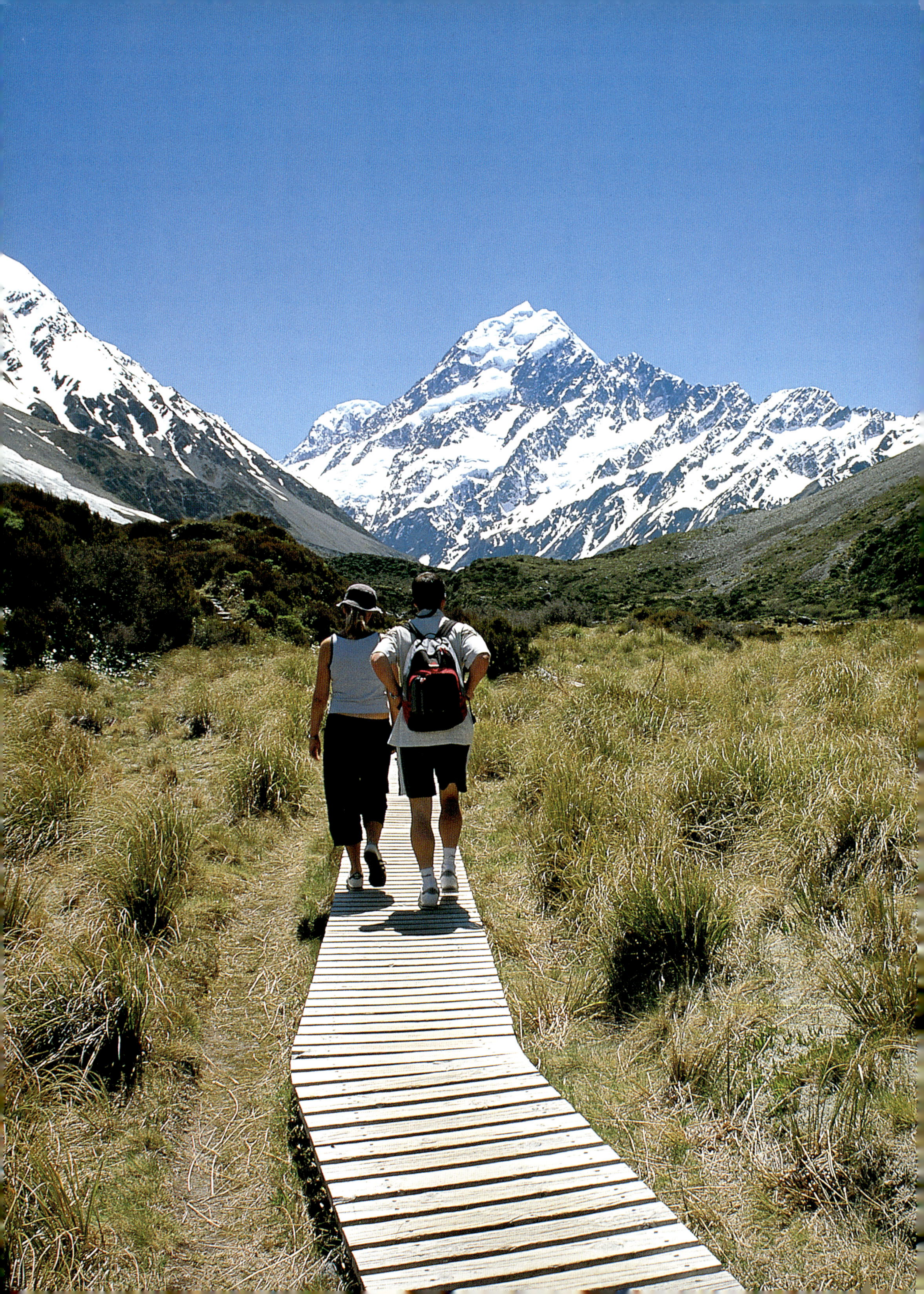

Opposite: Sensitive herbfields and grasslands are protected by a boardwalk along one of the most popular tracks in the national park. The Hooker Valley track leads from Mt Cook village, by way of the Alpine Memorial and two swing bridges, to Hooker Glacier.

Right: The eminent botanist Leonard Cockayne said of the Mount Cook lily: '. . . pure white, the petals at times so numerous that the flower looks like a semi-double . . . thirty blossoms to the stalk . . . what a glorious spectacle . . .'

Below: South from the village lies Governors Bush. An hour's easy walk passes through the green tranquillity of the only native forest in the national park where – in astonishing contrast to its alpine surroundings – temperate ferns grow among the gnarled boughs of silver beech heavily clothed with mosses, lichens and epiphytes.

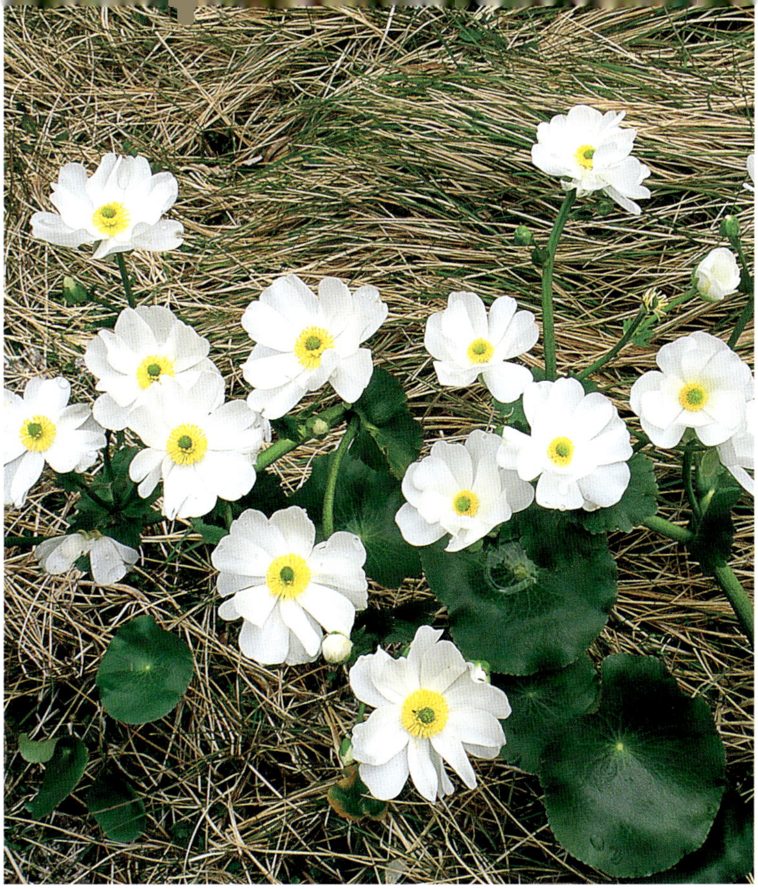

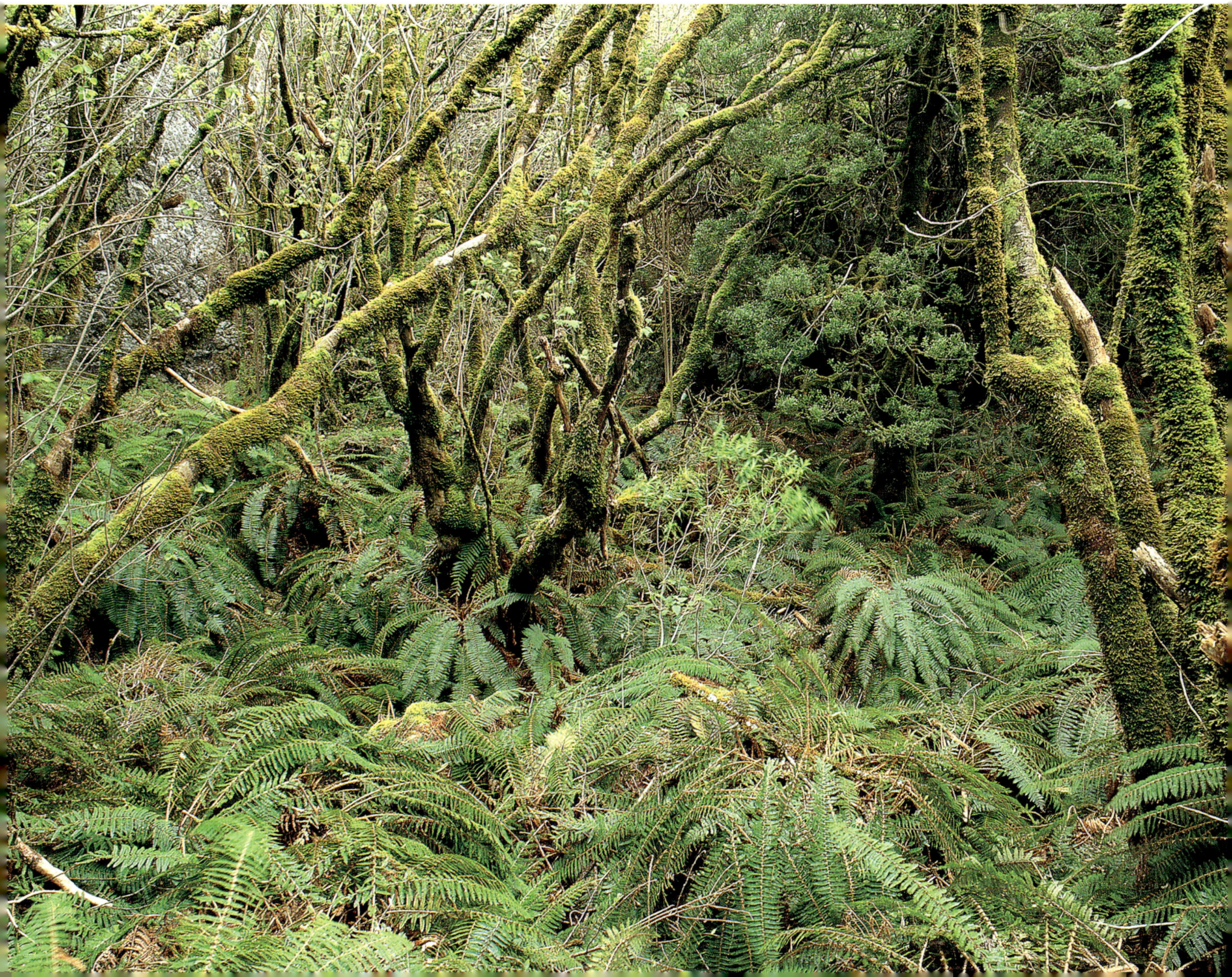

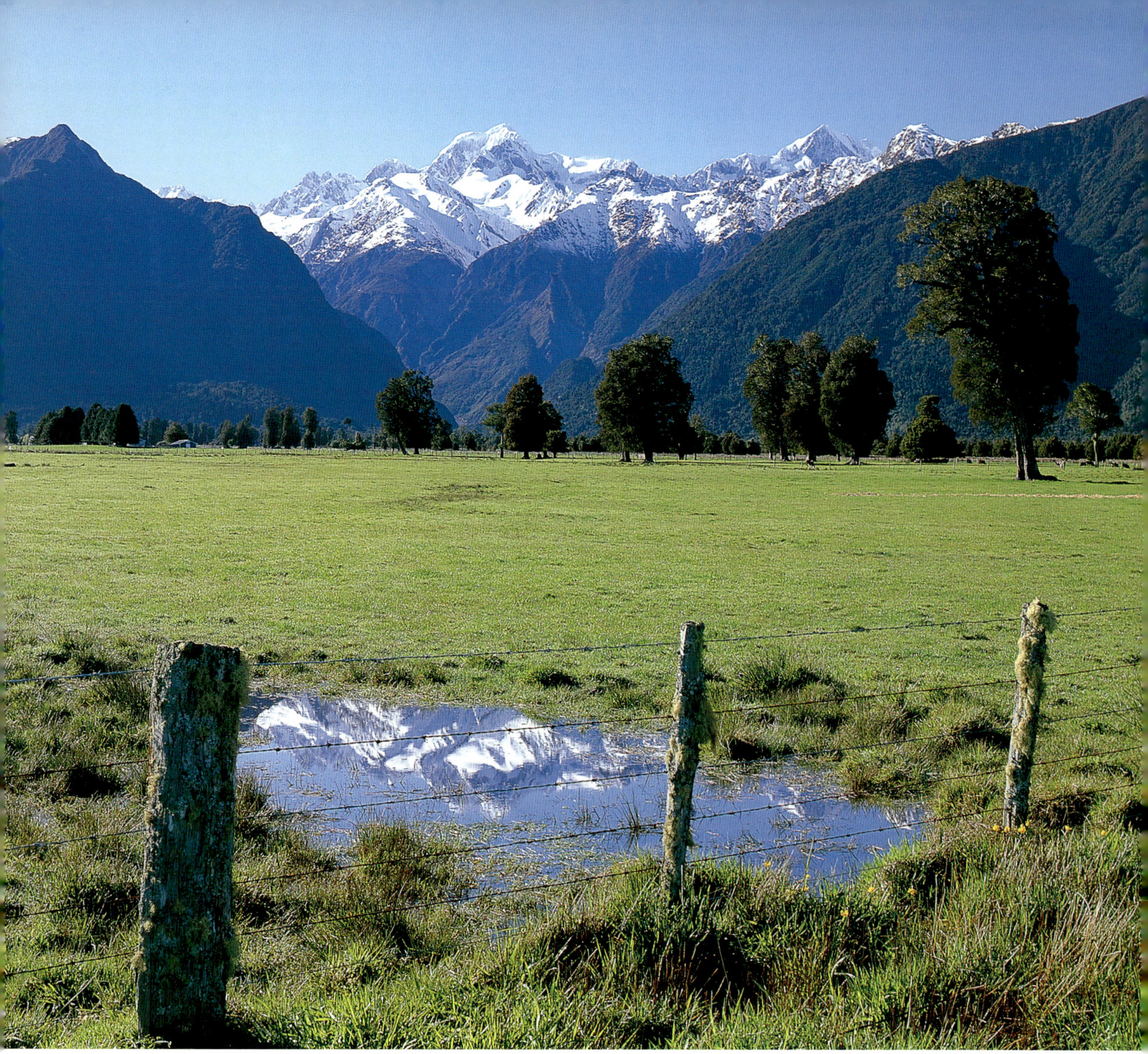

Above: A Westland puddle reflects Mt Tasman which, at 3496 m, approaches Aoraki/Mt Cook's height and certainly matches it for spectacle. H.E.L. Porter, a noted early mountaineer said of it: 'of all the mountains I have ever seen, in reality or in dreams, Tasman is the most faultlessly beautiful . . .'

Opposite above: Glentanner Station, a high country sheep run first taken up in 1857, lies on the east side of the rugged Ben Ohau Range beside the Tasman River. It is 25 km from Aoraki/Mt Cook village and offers tourist facilities: caravan and camp sites, cabins, horse-trekking, fishing and ski-flights to land on the glaciers.

Opposite below: There's been a Hermitage Hotel at Aoraki/Mt Cook since 1884, the first one built of clay and straw and accommodating only eleven guests. Today's hotel, the focal point for visitors, has 164 rooms with every desirable comfort; haute cuisine, and multilingual staff.

Overleaf: Sunrise reveals the three peaks of Aoraki/Mt Cook.

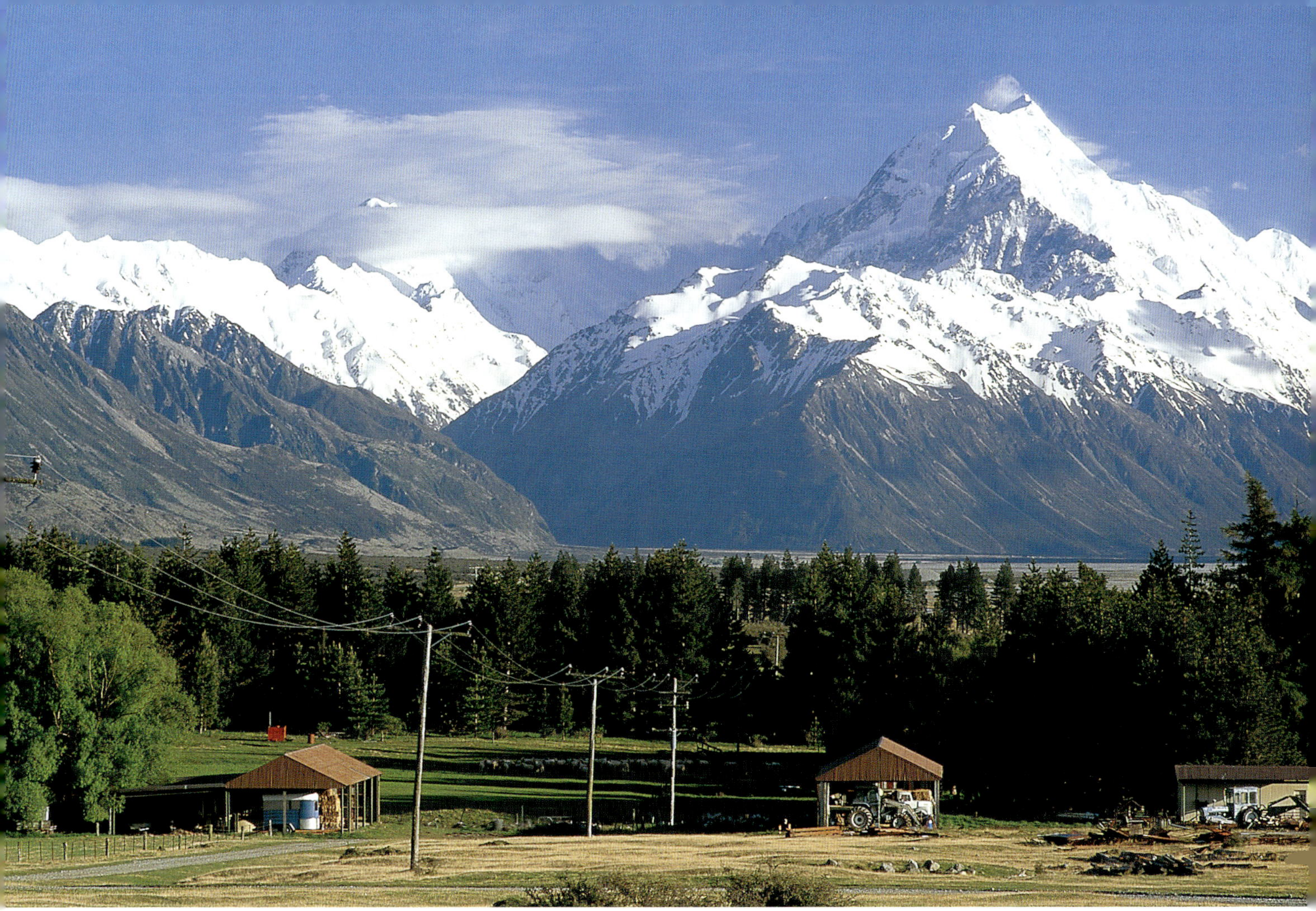

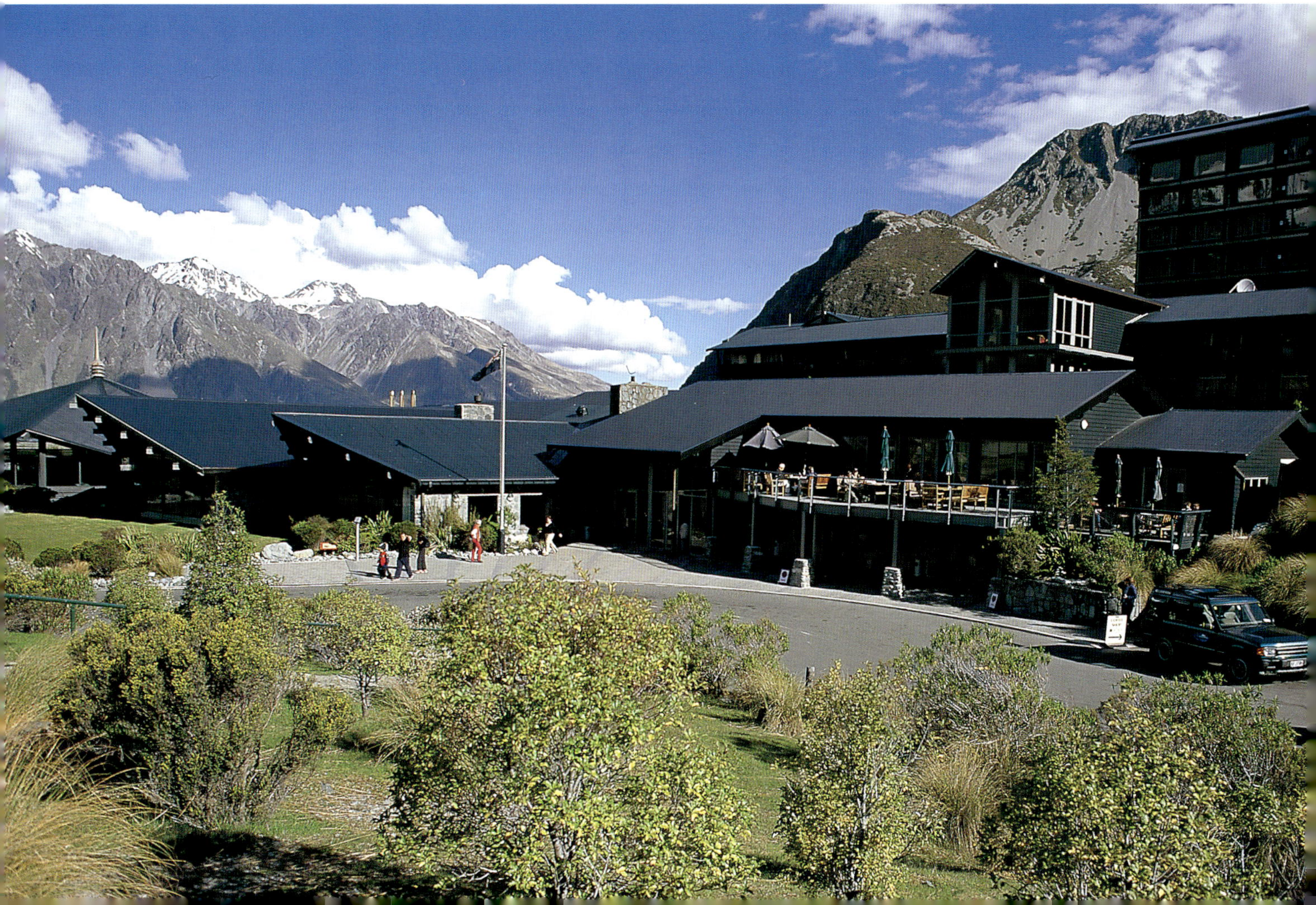

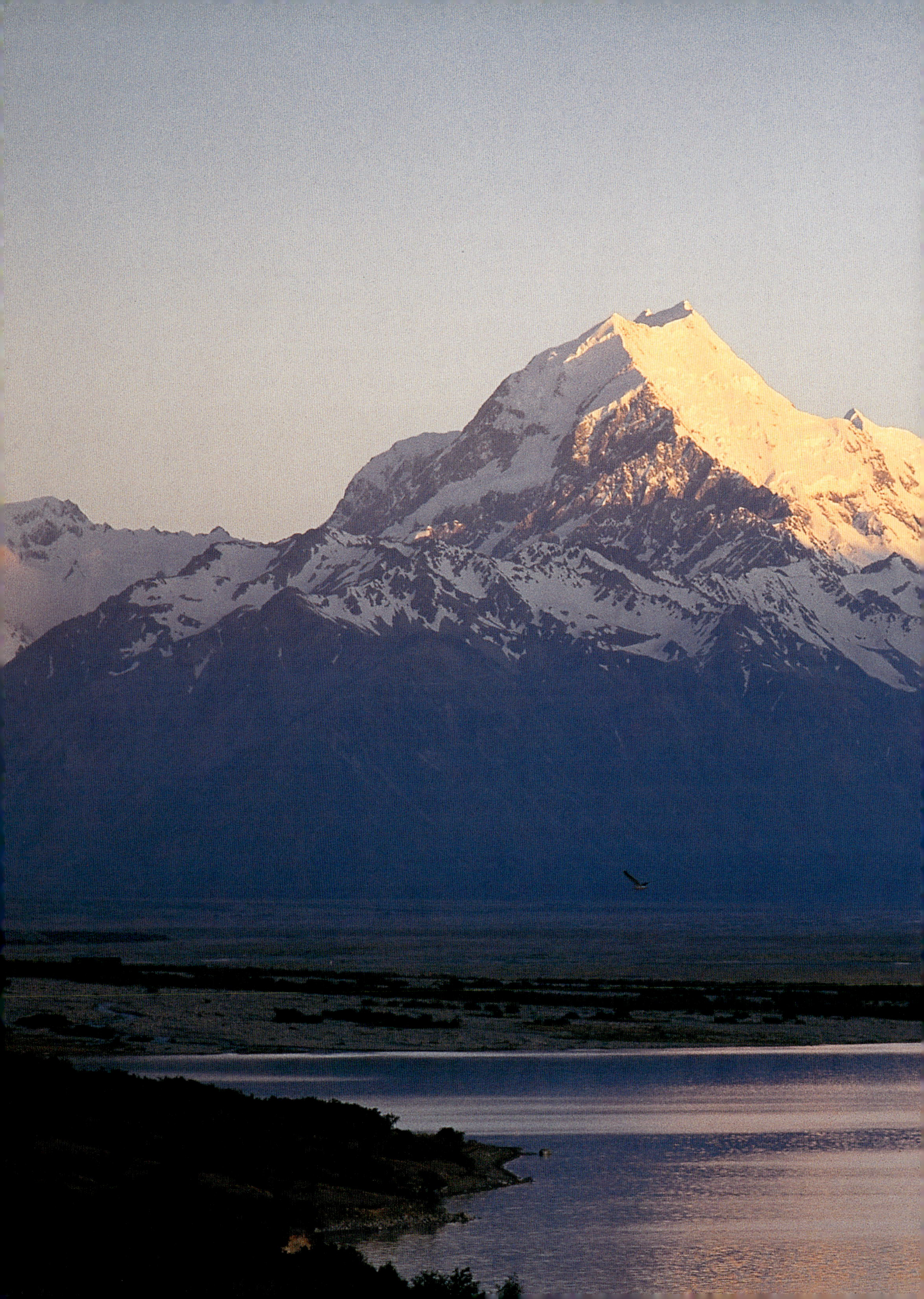

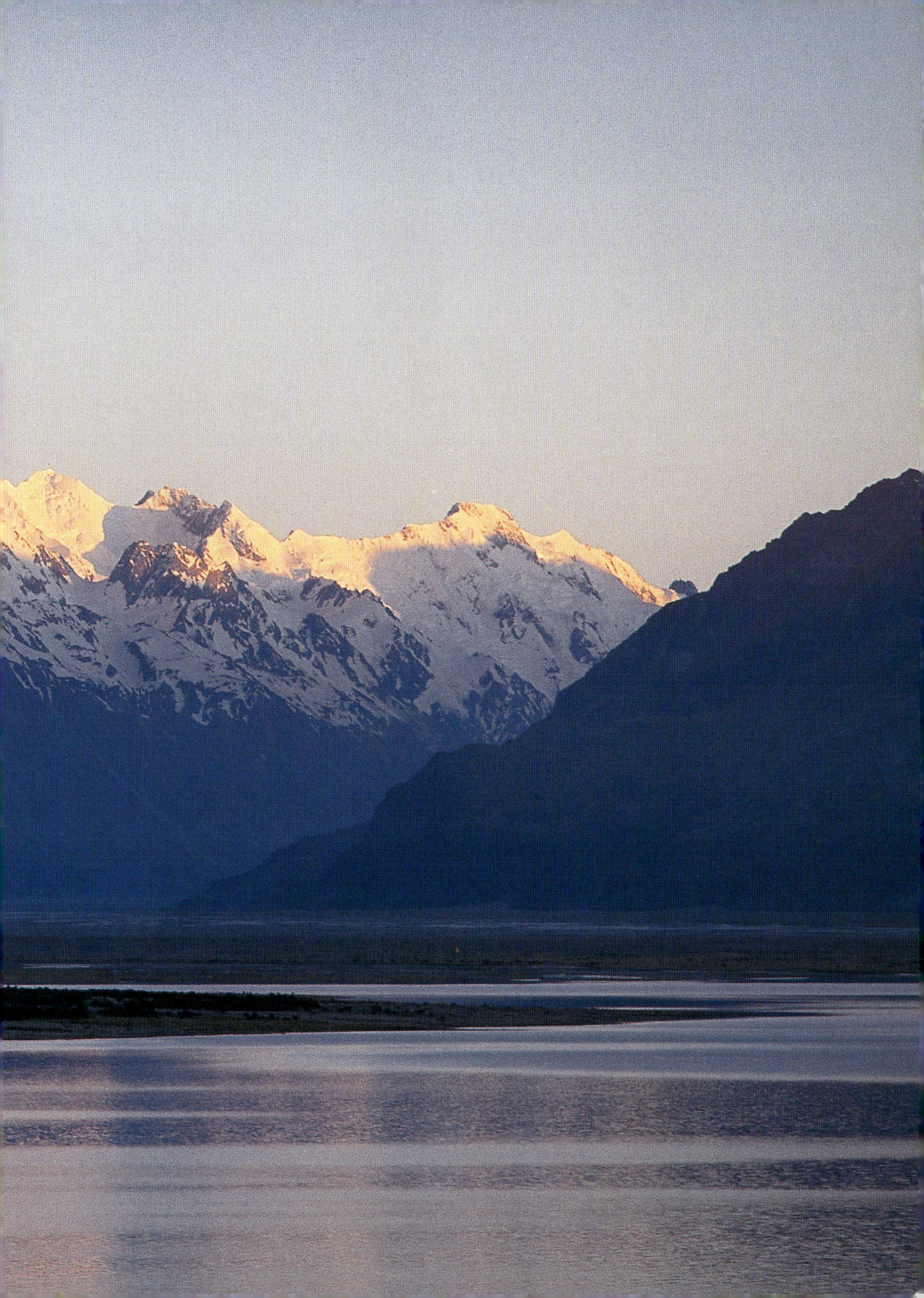

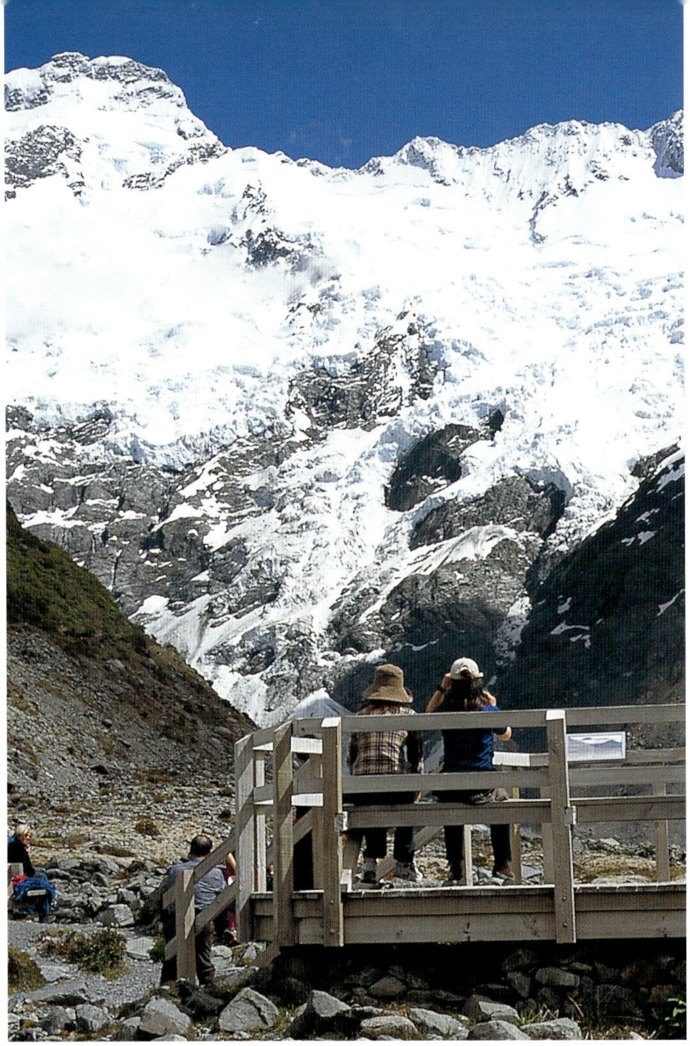
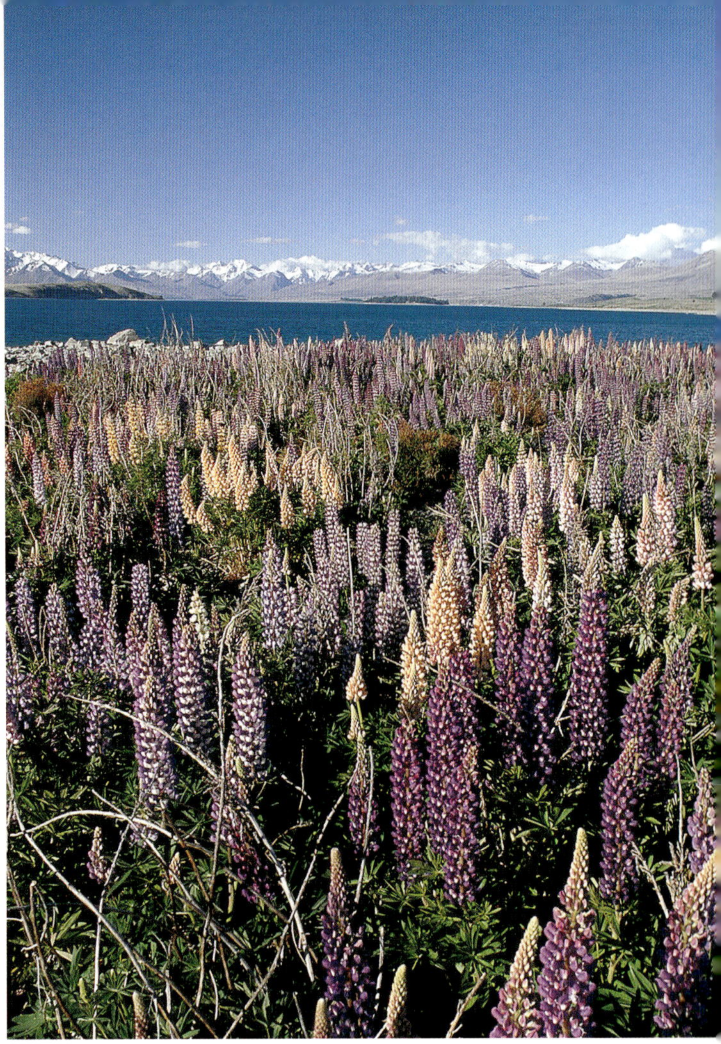
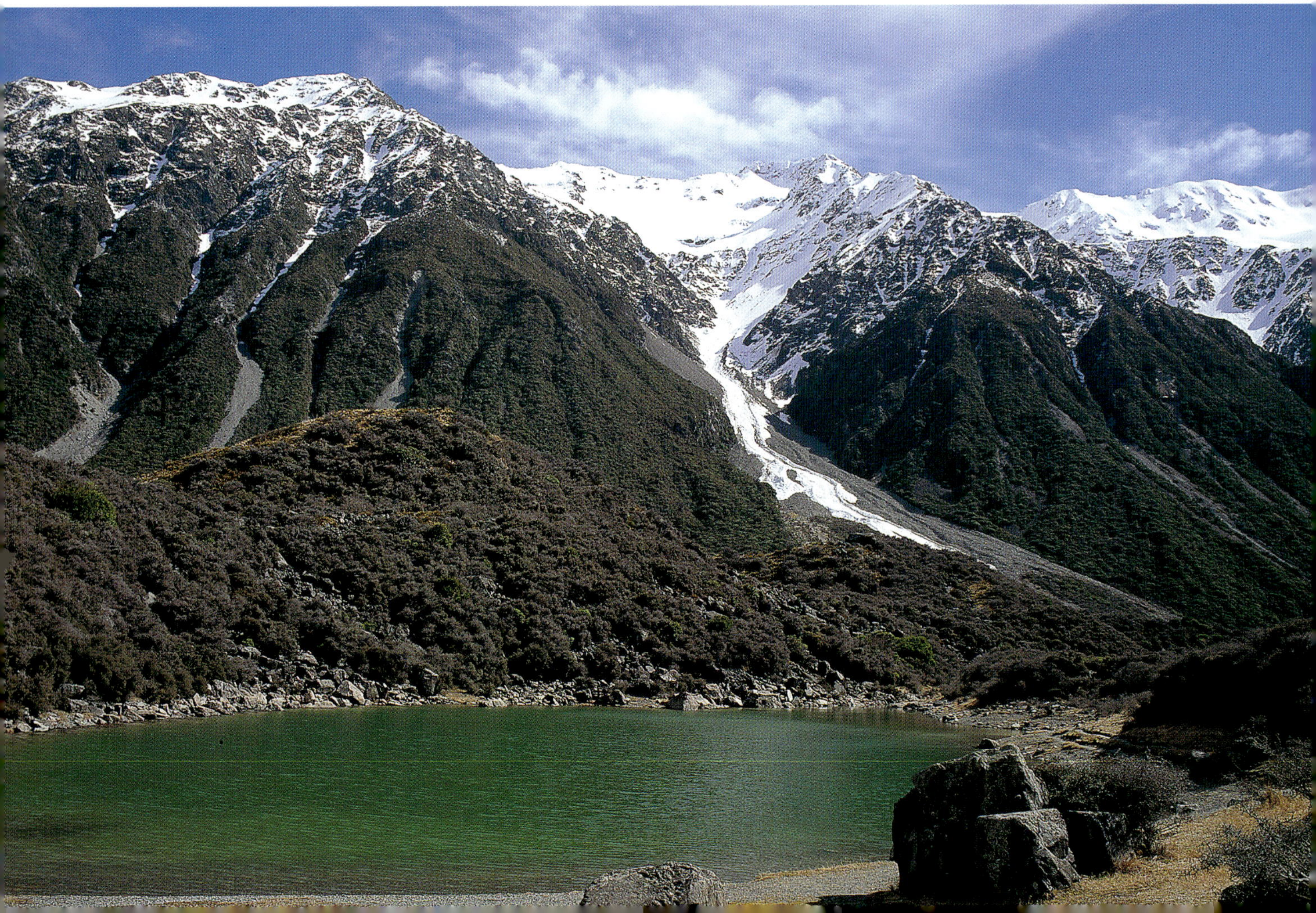

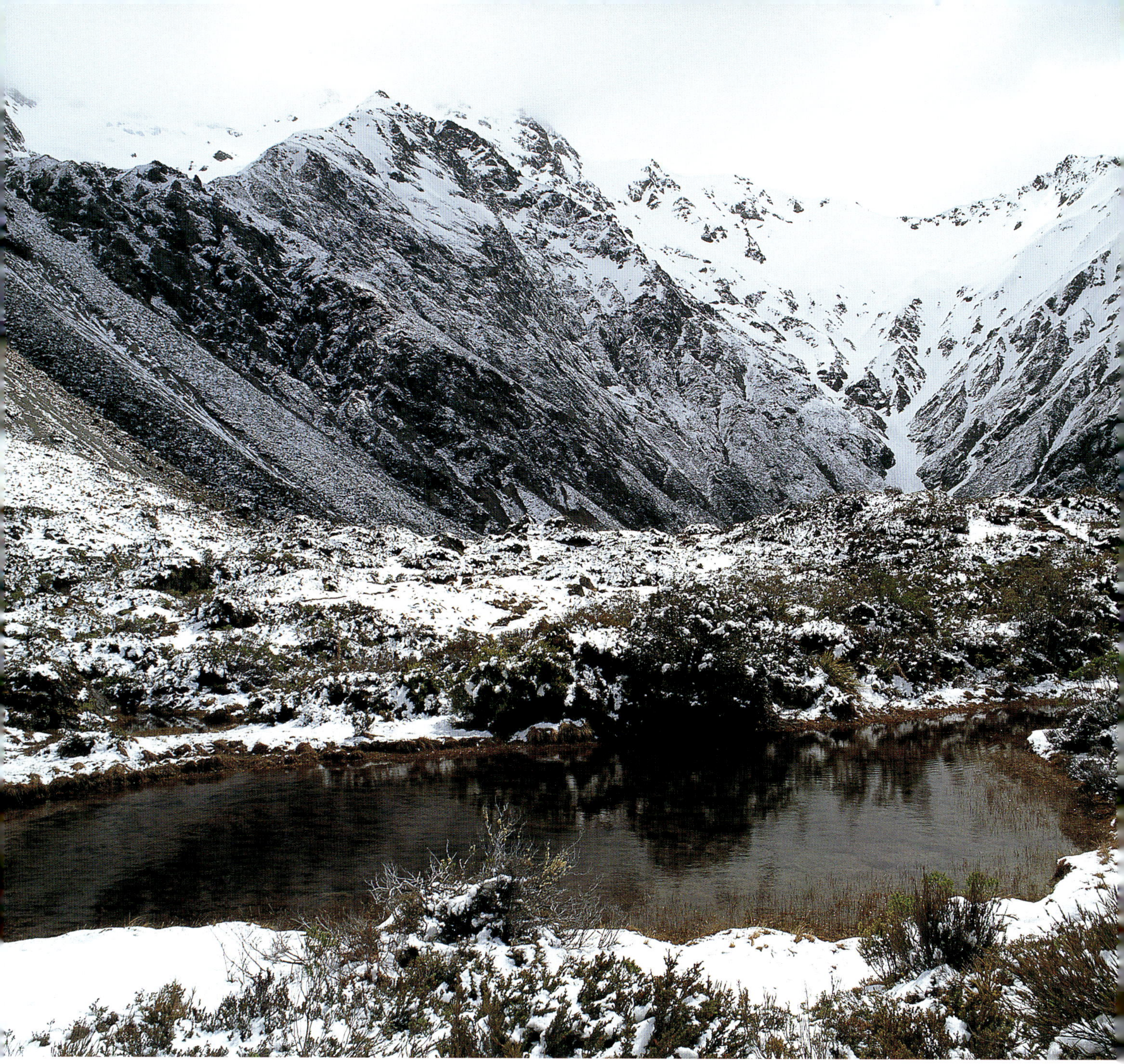

Opposite above left: From Kea Point, by easy track from the village across shingle fans and mountain scrub, walkers have a fine view across Mueller Glacier to the sheer snow face of Mt Sefton.

Opposite above right: The Godley Glacier which flows off the northern slopes of Mt D'Archiac (2865 m) feeds Lake Tekapo via the Godley River. At the southern end of the lake wild lupins grow profusely, splashing colour on an otherwise muted landscape.

Opposite below: On a track that runs up the west side of Tasman River to historic Ball Hut the Blue Lake – distinctly green – lies below the amphitheatre of Mt Wakefield.

Above: The Red Tarns – so called because of the colour of the weed that grows in their depths – are a short walk from the south end of Aoraki/Mt Cook village. They lie below Mt Sebastopol (1468 m) seen here shortly after a snowstorm.

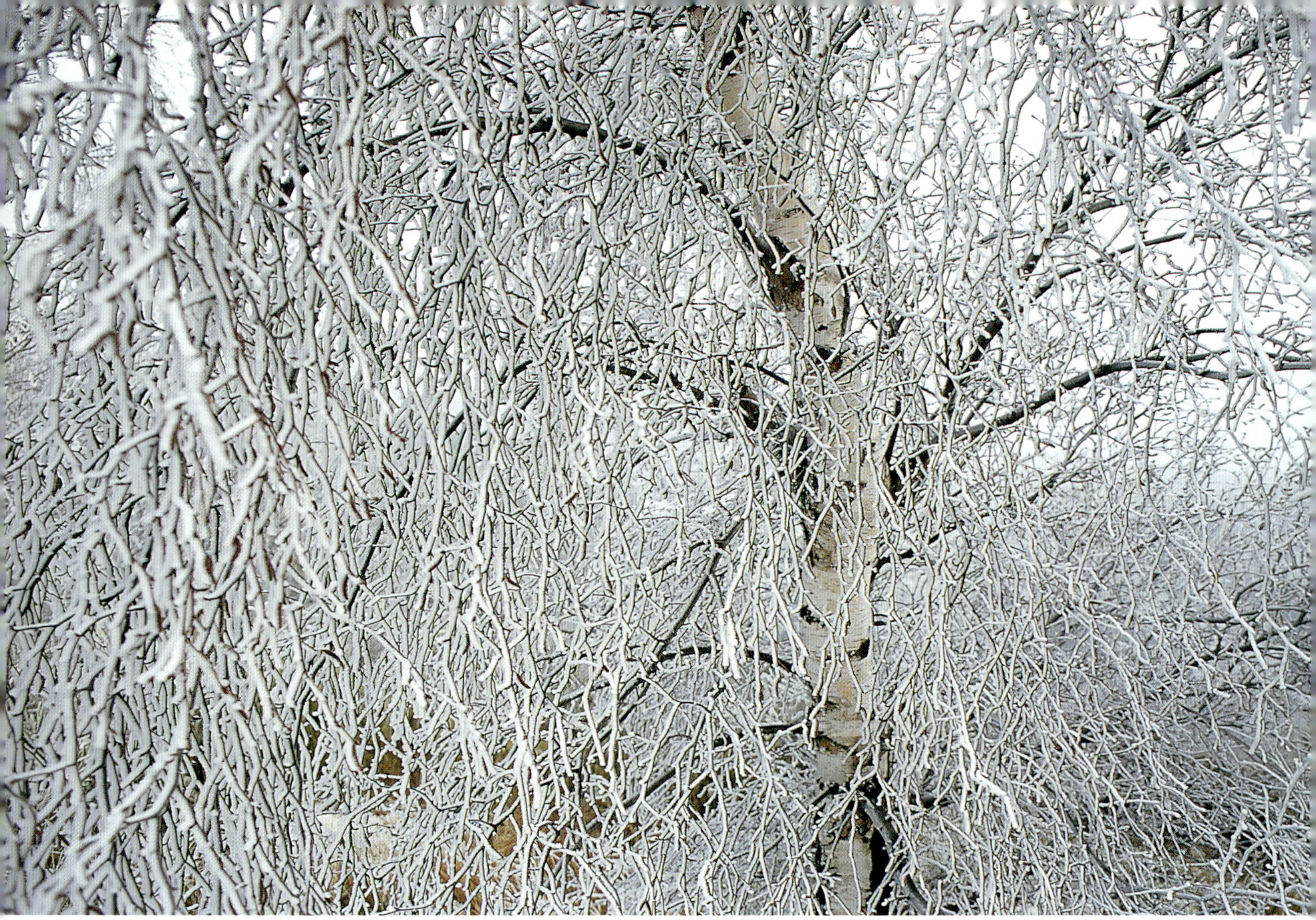

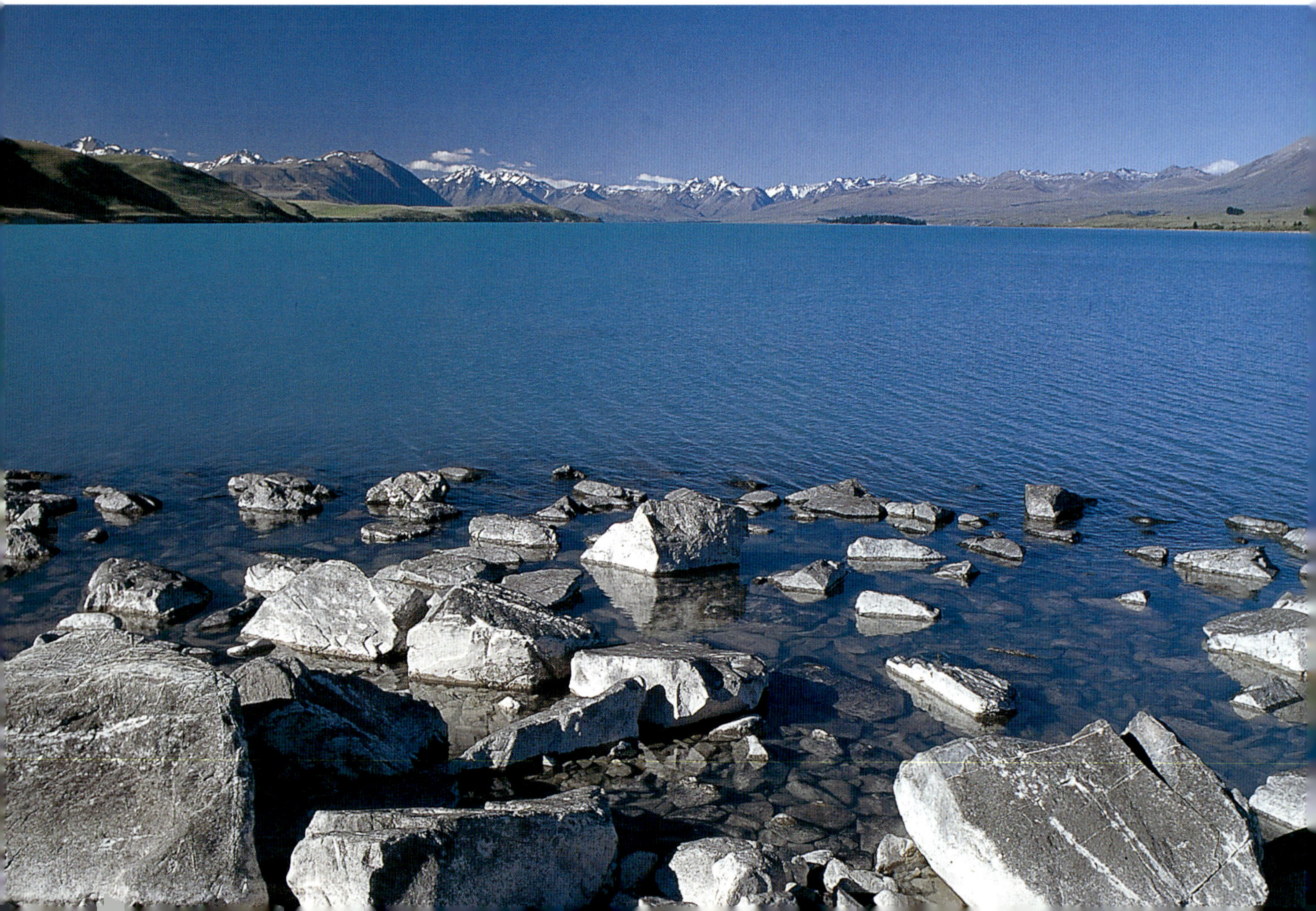

Opposite above: Intensely low temperatures lay a fine filigree of hoarfrost on a silver birch at Lake Pukaki.

Opposite below: Lake Tekapo, 20 km long, has its outlet controlled by a dam which is part of the extensive Upper Waitaki Power Scheme. Thus the lake's level may fluctuate between 710 and 704 metres, at its lowest revealing a litter of rocks deposited by ancient glacial activity.

Right: Often misrecorded as being the dog of McKenzie, an alleged sheep stealer, this bronze statue, sculpted by a Mackenzie Country woman Mrs Innes Elliot and unveiled in 1968, is a general tribute to the sheepdog, '. . . without the help of which, the grazing of this mountain country would be impossible.'

Below: Not far from the sheepdog statue at the lower end of Lake Tekapo stands the Church of the Good Shepherd. Its foundation stone was laid by the Duke of Gloucester in 1935 and it was finished later that year as a memorial to the settlers and shepherds whose sheep ranged across wild Mackenzie Country. It is built of undressed lakeside boulders.

Overleaf: Distant Aoraki/Mt Cook across Lake Pukaki.

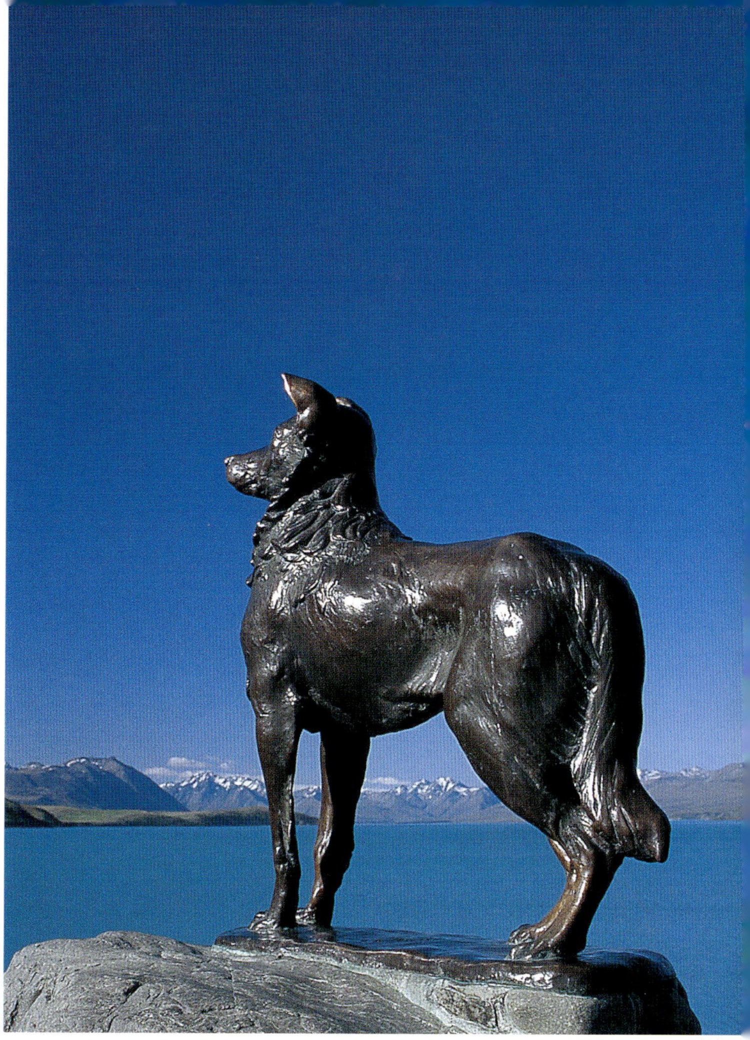

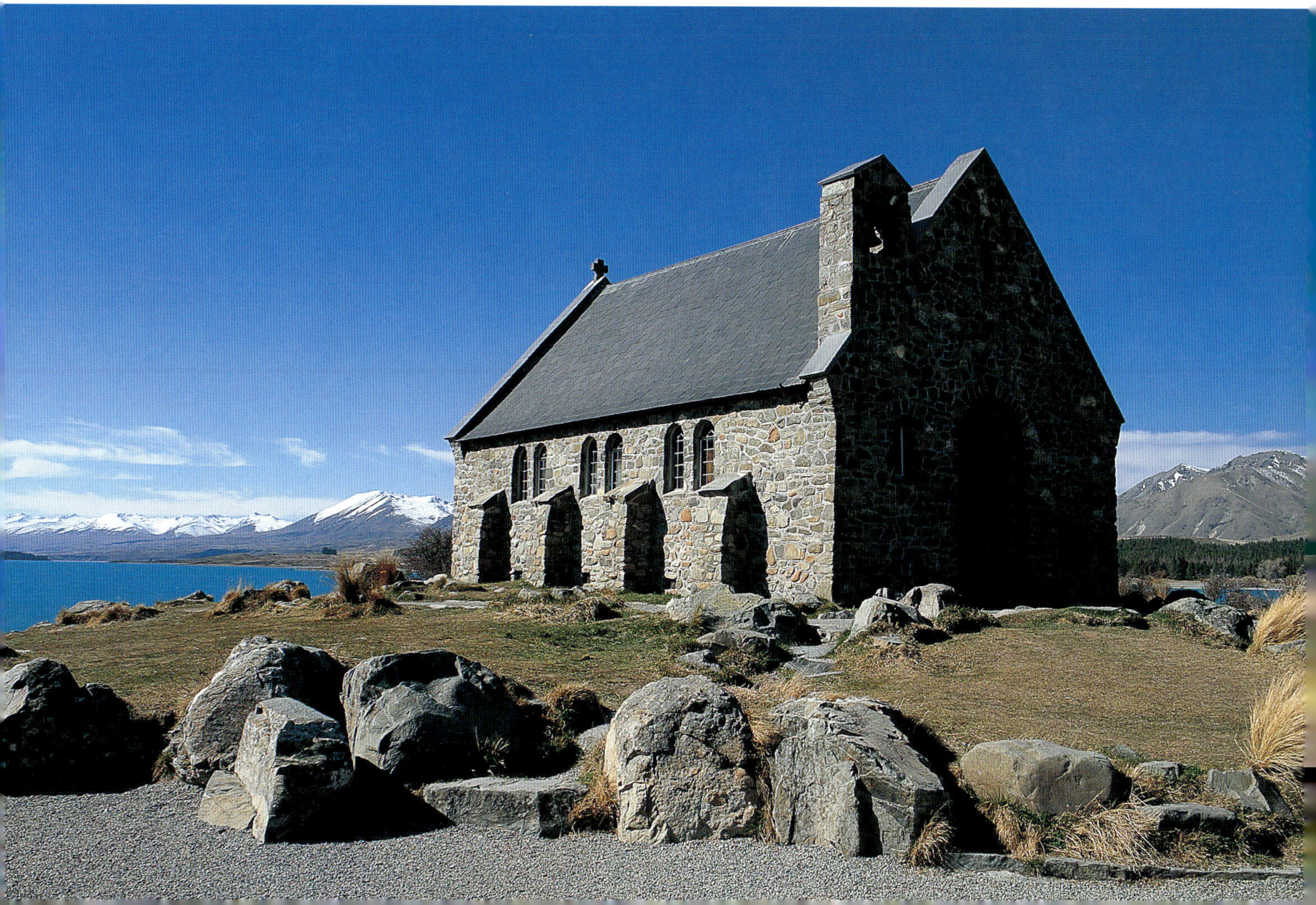

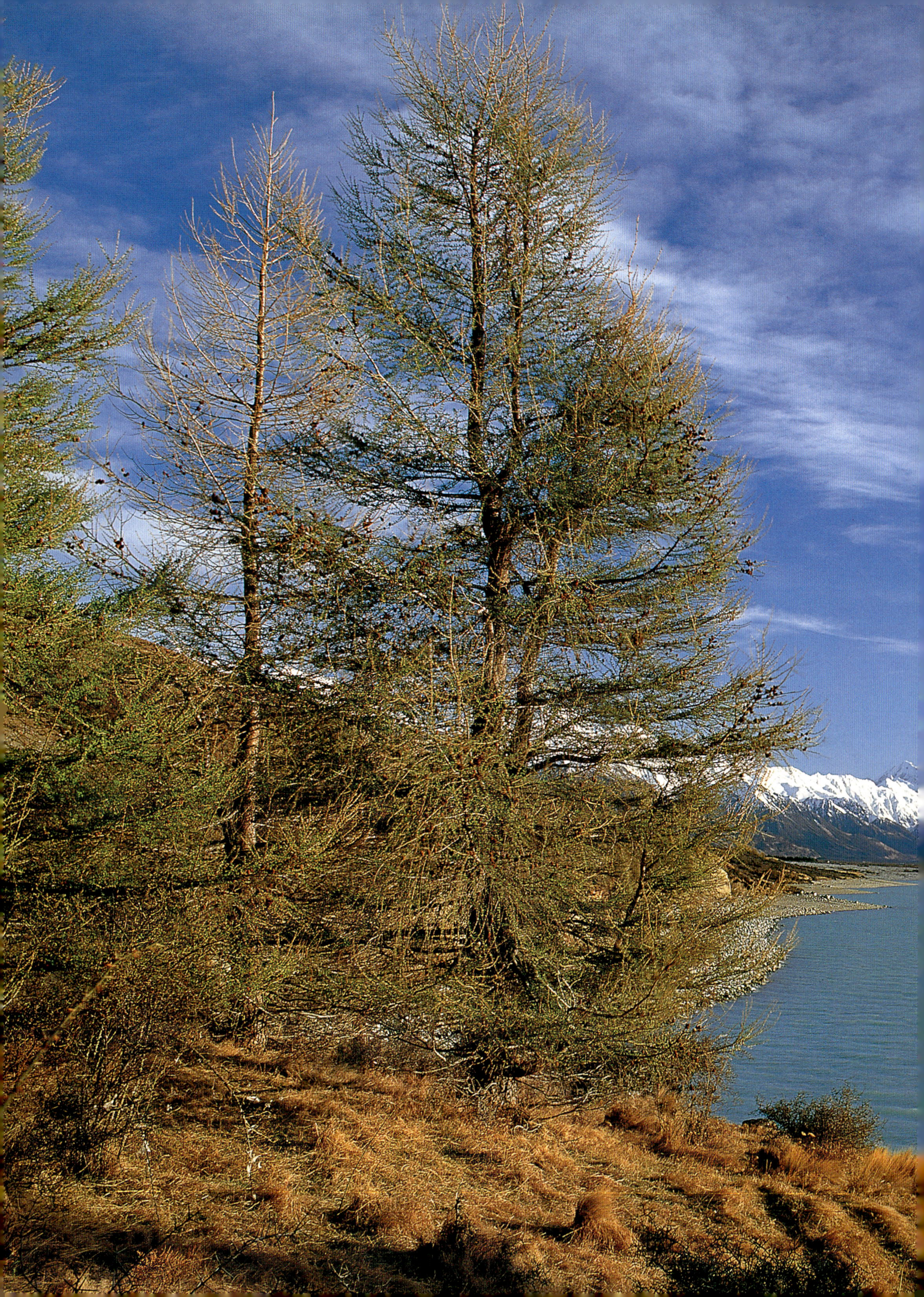

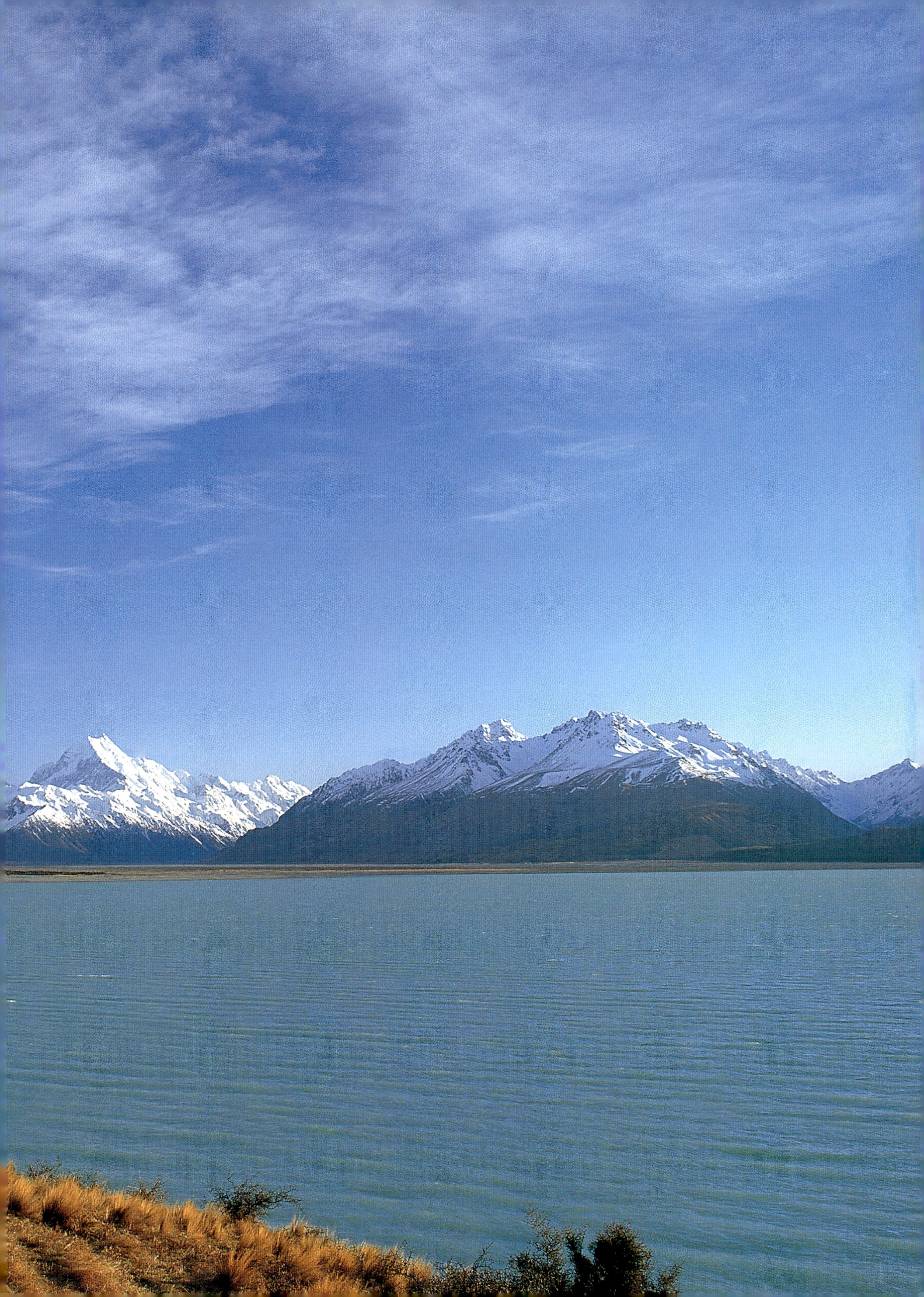

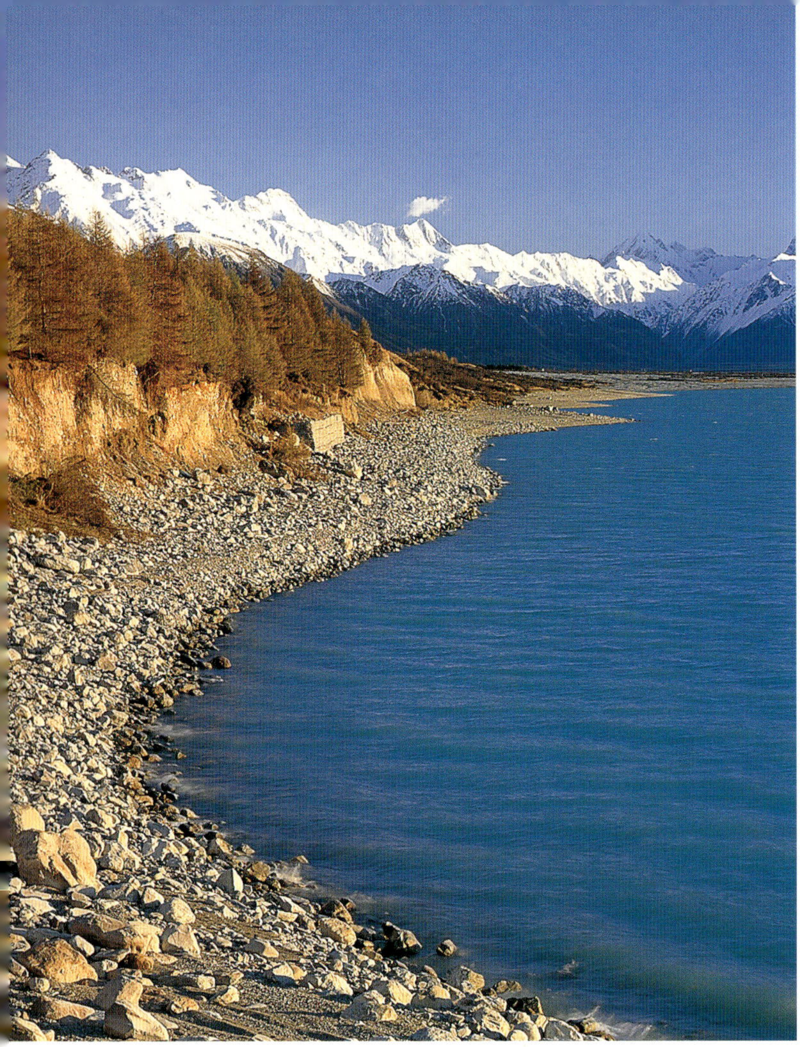

Left: Exotic pines border the low cliffs of Lake Pukaki.

Below: At Burke Pass, where inland Canterbury gives way to Mackenzie Country, an old weatherboard hut quietly struggles to survive against nature and the elements. On many a snowed-in night its sheltering walls and roaring fire would have made the difference between life and death for shepherds working the hill flocks of the nearby Two Thumb Range.

Opposite above: The controlled waters of the mighty rivers emerging from Lakes Tekapo, Pukaki and Ohau feed into artificial Lake Benmore, highest of a series of hydro-electric power lakes along the Waitaki River that forms the boundary between the provinces of Canterbury and Otago.

Opposite below: The prairie-like slopes of the alpine region are laced with little streams; cool, clear and fast, often home to surprisingly large trout. This typical burn drains the Wild Mans Brother Range and flows into Lake Heron.

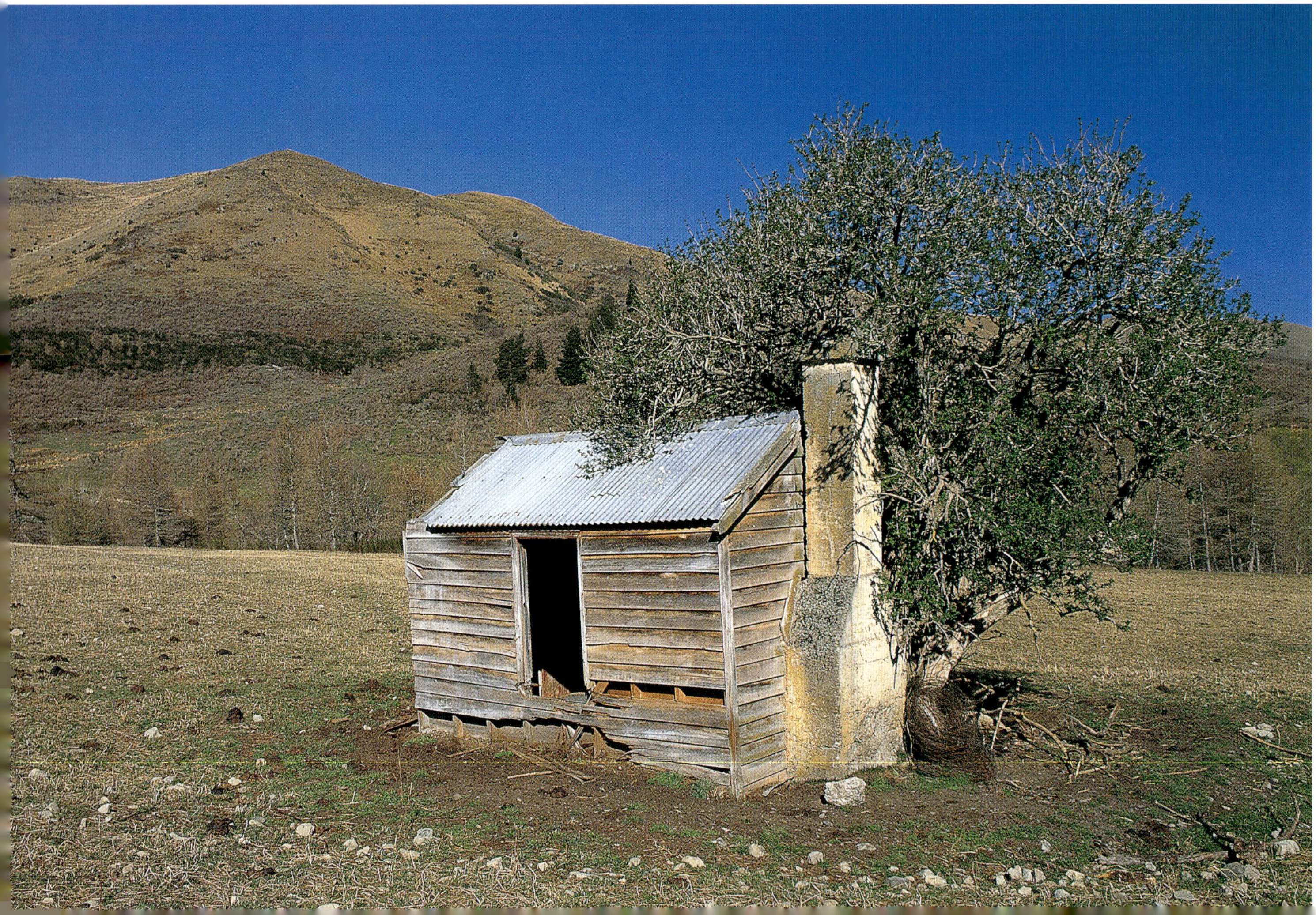

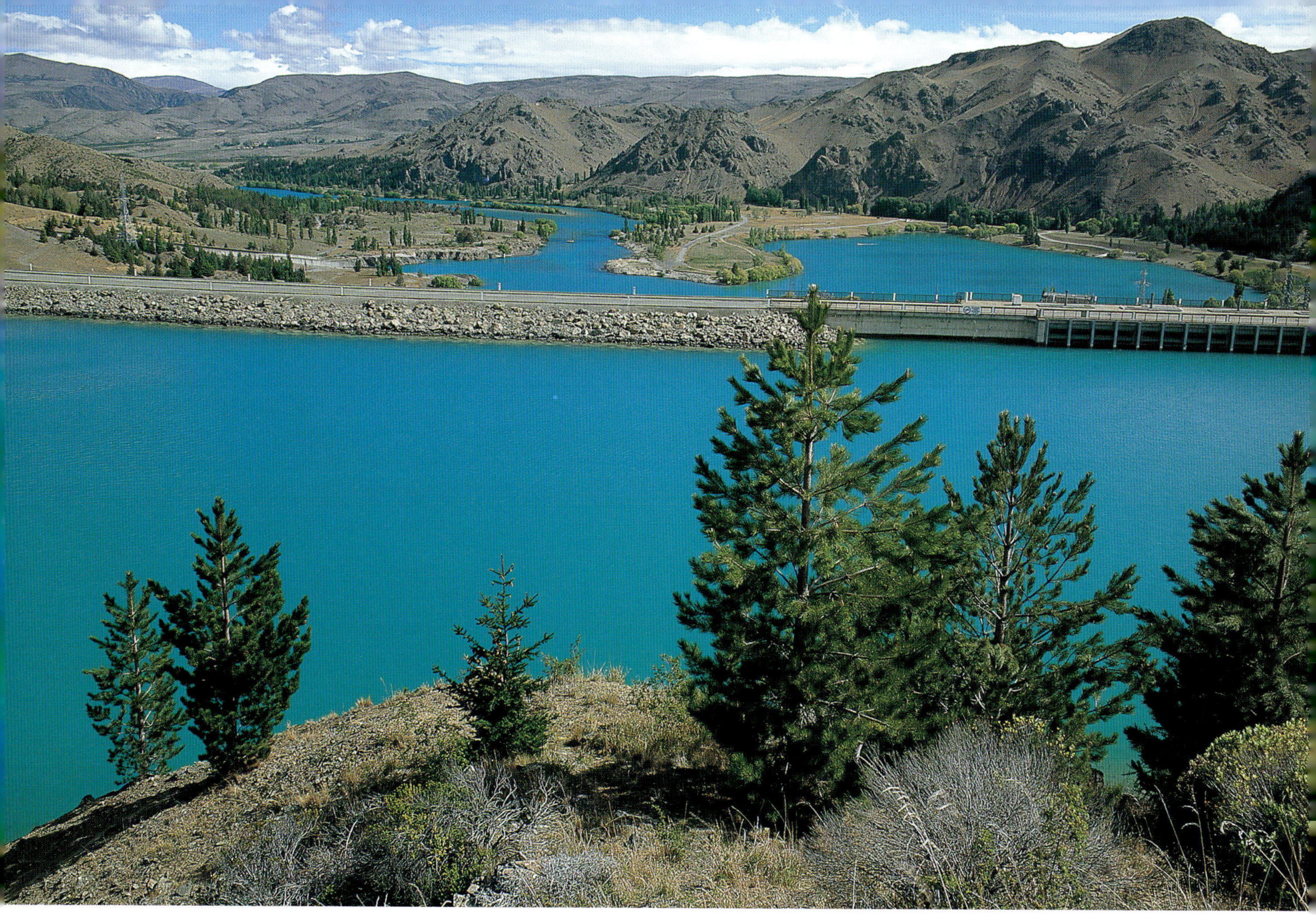

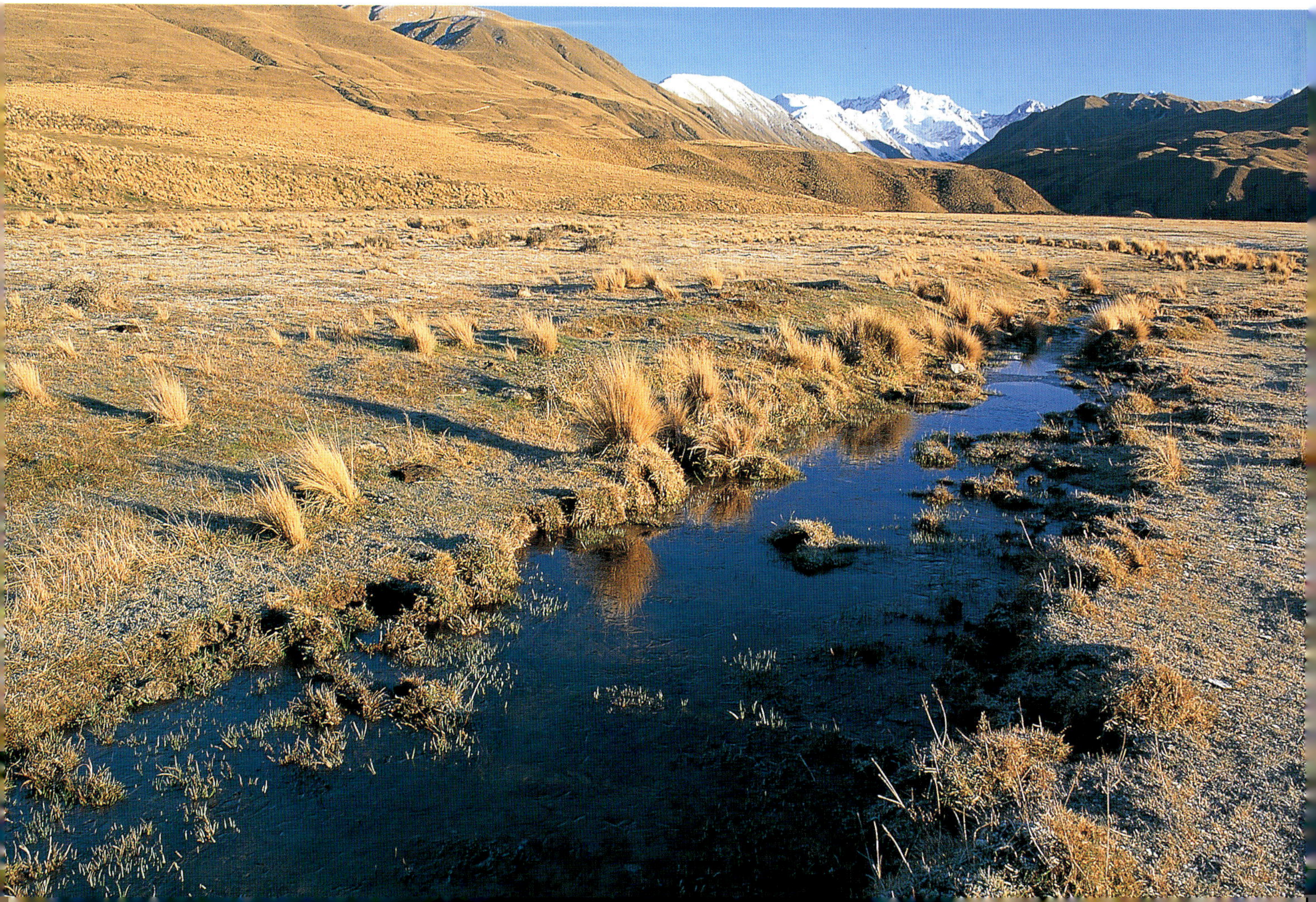

Right: Cathedral-like weathered pinnacles rise in the valley of the Ahuriri River which runs north of Omarama and into the west arm of Lake Benmore. The cliffs of layered silt and clay change colour depending upon the light and time of day.

Below: At the southern reaches of alpine influence the Lindis Pass leads from Mackenzie Country into inland North Otago. On rare occasions motorists will patiently forego speed for the reward of a truly rural sight of driven sheep grazing the roadside berms.

Opposite above: The village of Omarama lies strategically on access roads to the alpine region, Canterbury, coastal Otago and Central Otago. At their junction this statue, sculpted by Bill Adams, pays tribute to the merino sheep that have grazed the high hills and brought wealth to the district since the nineteenth century.

Opposite below: Omarama is an international gliding centre where world records have been set. Sail-planes flying within the mountain-rimmed Mackenzie Basin can take advantage of nor-westerly updraughts that may rise to 15 000 metres.

Overleaf: Fresh snowfall near the summit of Lindis Pass.

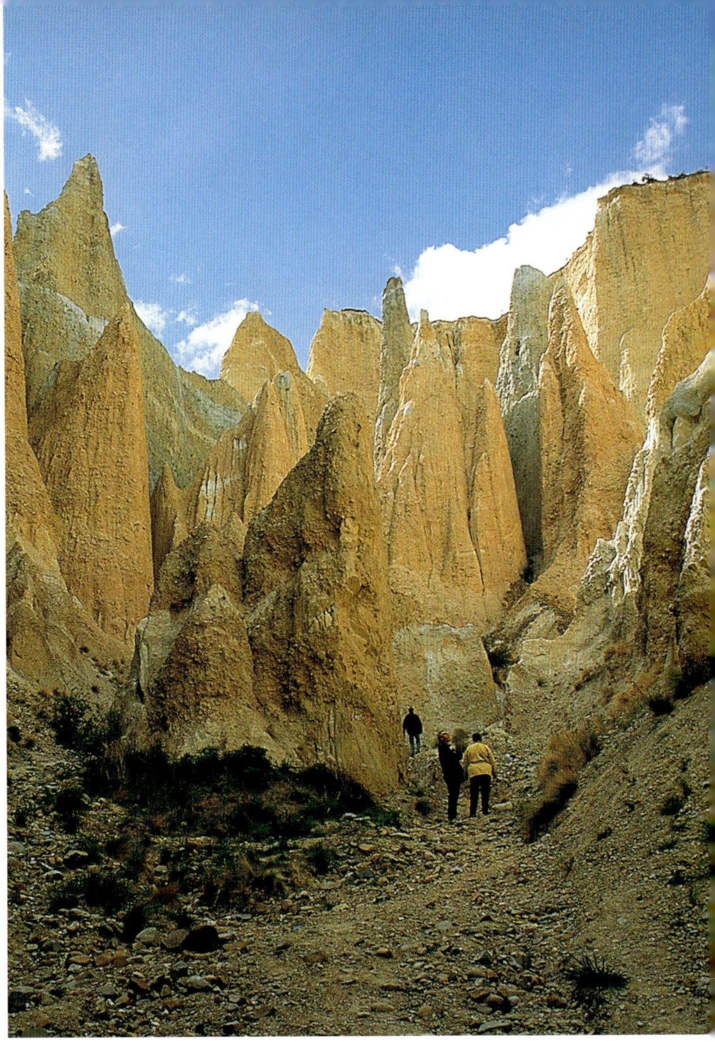

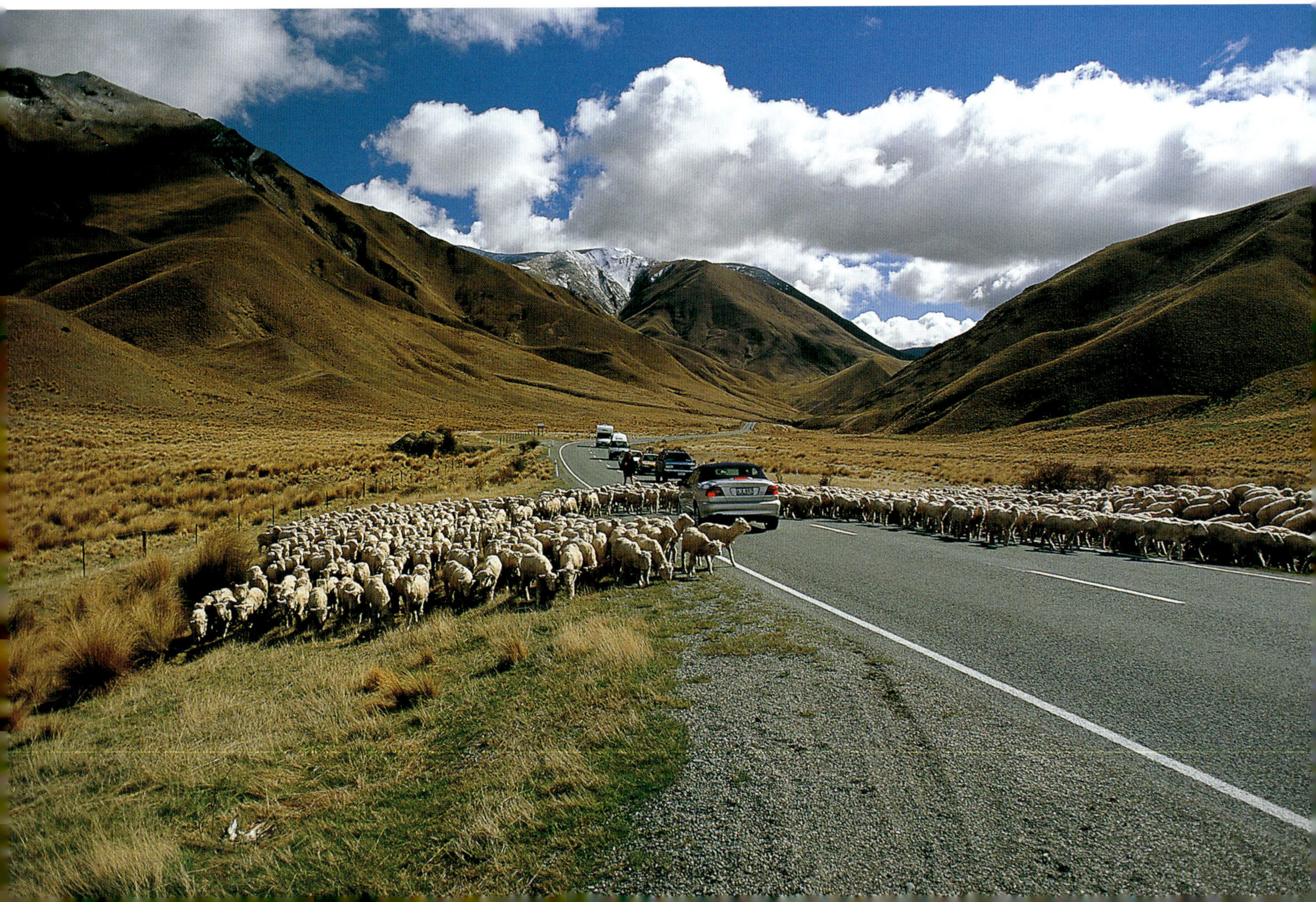

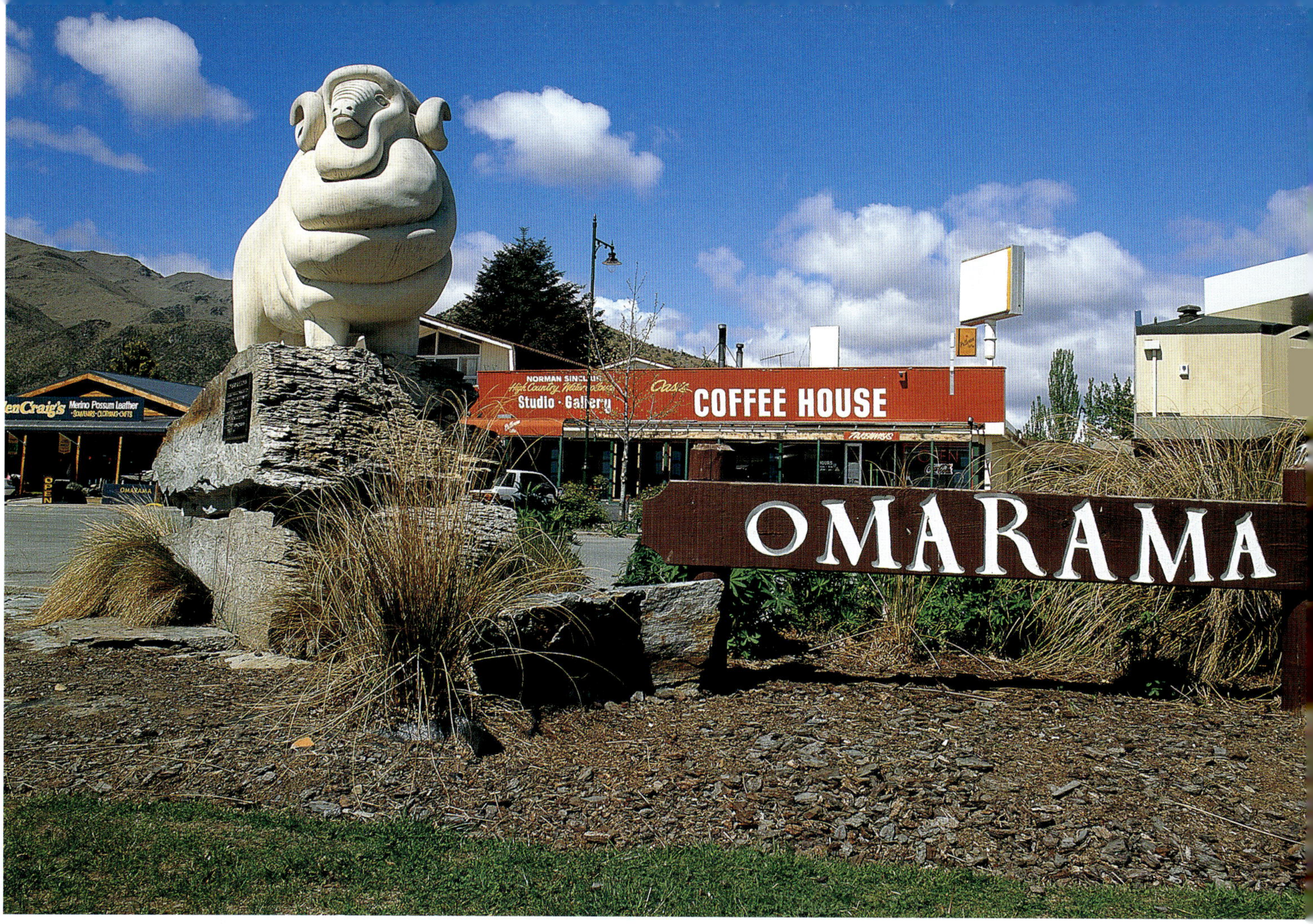
Studio · Gallery
COFFEE HOUSE
OMARAMA

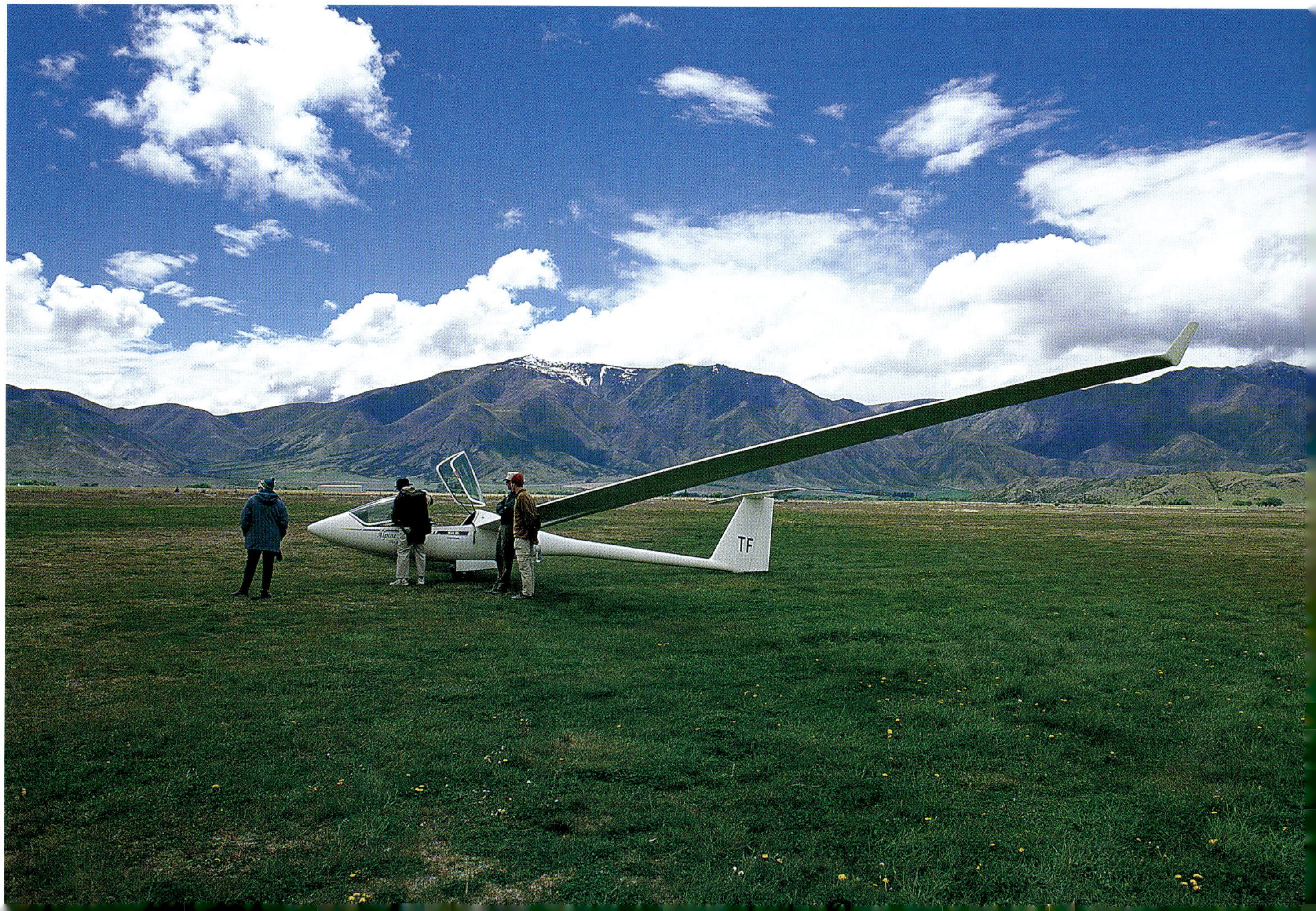
TF

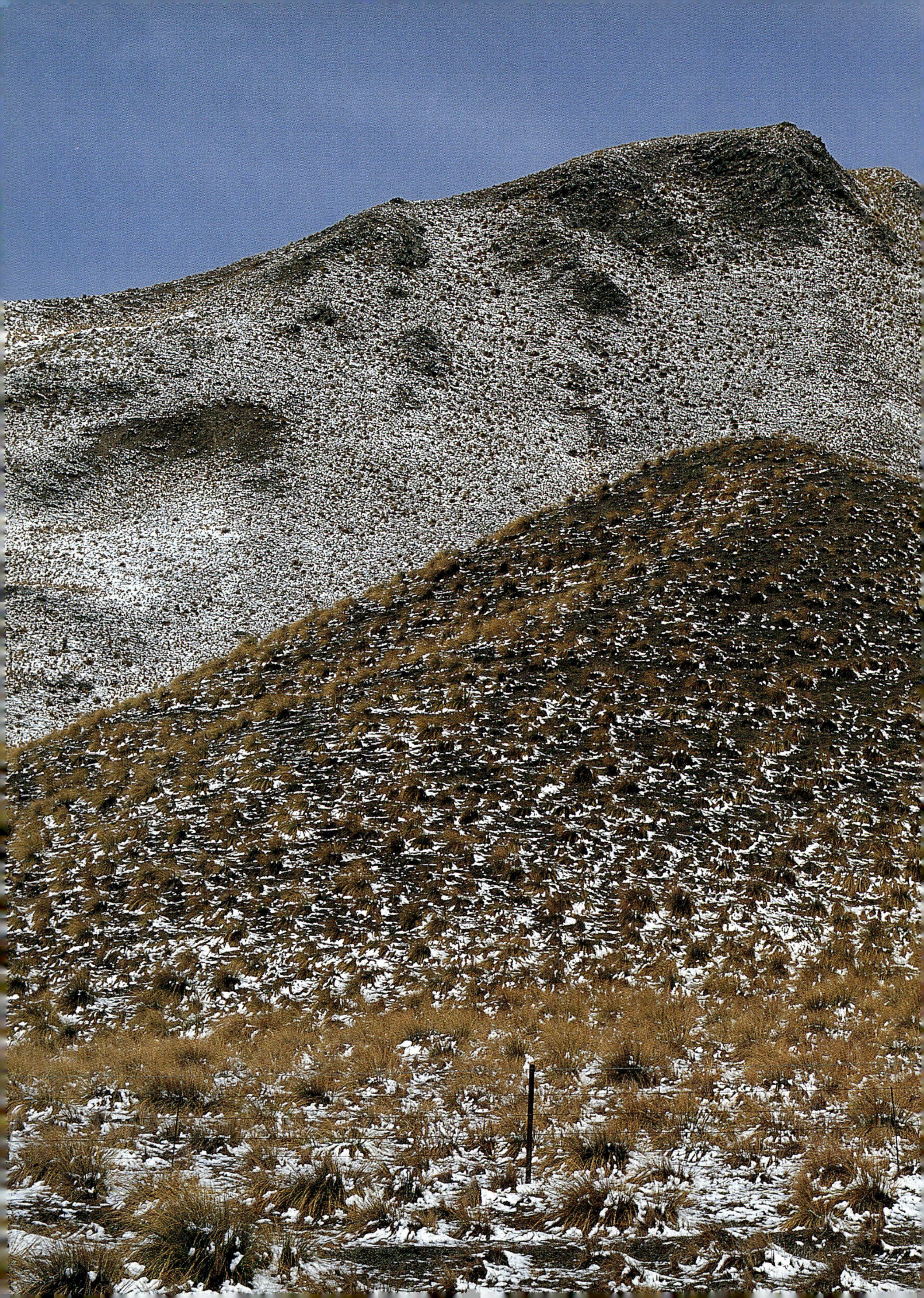